Optical System Design

ISBN 0-13-901042-4

90000

9 780139 010422

Optical System Design

Allen Nussbaum
Electrical Engineering Department
University of Minnesota

To join a Prentice Hall PTR
mailing list, point to:
http://www.prenhall.com/mail_lists/

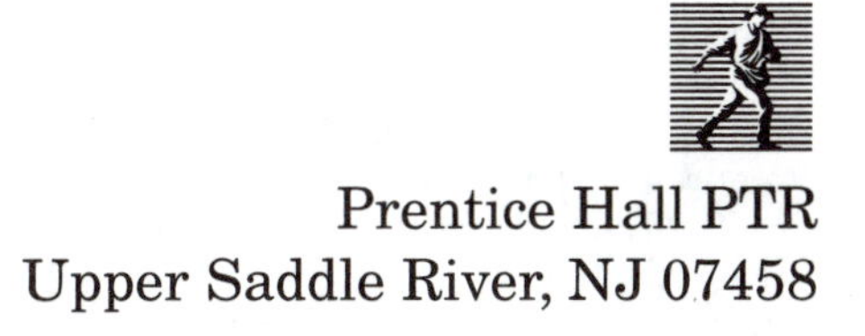

Prentice Hall PTR
Upper Saddle River, NJ 07458

Library of Congress Cataloging-in-Publication Data

Nussbaum, Allen.
Optical system design / Allen Nussbaum.
p. cm.
Includes index.
ISBN 0-13-901042-4 (case)
1. Geometrical optics. 2. Optical instruments—Design and construction. 3. Matrices. I. Title.
QC381.N85 1998
681'.4—dc21 97-43430
CIP

Editorial/production supervision: *BooksCraft, Inc., Indianapolis, IN*
Cover design director: *Jerry Votta*
Cover design: *Alamini Designs*
Acquisitions editor: *Bernard M. Goodwin*
Manufacturing manager: *Alexis R. Heydt*
Marketing manager: *Betsy Carey*

Prentice Hall books are widely used by corporations and government agencies for training, marketing, and resale.

The publisher offers discounts on this book when ordered in bulk quantities. For more information, contact: Corporate Sales Department
Phone: 800-382-3419
Fax: 201-236-7141
E-mail: corpsales@prenhall.com

Or write: Prentice Hall PTR
Corp. Sales Dept.
One Lake Street
Upper Saddle River, NJ 07458

Printed in the United States of America

10 9 8 7 6 5 4 3 2 1

ISBN: 0-13-901042-4

Prentice-Hall International (UK) Limited, *London*
Prentice-Hall of Australia Pty. Limited, *Sydney*
Prentice-Hall Canada Inc., *Toronto*
Prentice-Hall Hispanoamericana, S.A., *Mexico*
Prentice-Hall of India Private Limited, *New Delhi*
Prentice-Hall of Japan, Inc., *Tokyo*
Simon & Schuster Asia Pte. Ltd., *Singapore*
Editora Prentice-Hall do Brasil, Ltda., *Rio de Janeiro*

To my wife

Barbara

Where are the words to describe more than 50 years of constant love?

Contents

Preface

Why another book on geometrical optics when there are so many already available? The principal reason is that the teaching of geometrical optics, both elementary and advanced, needs to be modernized. Two recent developments have made this possible. The first is the ready availability of information on the use of matrices. Ray tracing can be made simple through the use of 2 x 2 matrices, a method developed many years ago in England independently by R. A. Sampson and by T. Smith. Their ideas were made known by W. Brouwer in his published lecture notes, "Matrix Methods in Optical Instrument Design," W. A. Benjamin (1964). The second big change in optics education involves the use of personal computers; the programs given here will eliminate much of the complicated algebra in traditional courses. The amount of physics used is minimal: just Snell's law, Huygens' principle, and some wave theory. The emphasis is on geometry and the associated calculations.

Another area that needed improvement in teaching is Fourier optics. Here, unfortunately, there is no simplifying tool like matrices. Instead, it has become possible to eliminate many of the advanced topics in Fourier analysis and achieve a useful presentation again by taking advantage of the power of programming. It should also be pointed out that there are a moderate number of misconceptions about aberration theory in many of the well-known optics texts, and I have explained these in some detail. It is my hope that this book will encourage the use of new ways to study advanced geometrical optics so that it can be an enjoyable and rewarding experience.

Allen Nussbaum
Minneapolis, MN
nussbaum@ee.umn.edu

Acknowledgments

The text and equations in this book were originally produced with WordPerfect 5.1. The 3 1/4" x 3 1/4" illustrations were drawn with CoDraw, a product of CoHort Software, PO Box 19272, Minneapolis, MN 55419.

CHAPTER 1

Paraxial Matrix Optics

1.1 The Newton and Gauss Lens Equations

A simple derivation of the lens equations can be based on two experimental observations. The first is well known and is pictured in Figure 1.1. Parallel rays of light passing in air through a double convex lens (a simple magnifying glass, for example)—and not too far from the central ray—meet at the *focal point* or *focus*, designated as F′. For rays from the other side of the lens, another such point, F, exists that is the same distance from the lens as F′.

The second experiment uses the lens to magnify (Figure 1.2a); the image is larger than the object and oriented the same way. The amount of magnification decreases as the lens is brought closer to "ISAAC NEWTON" (Figure 1.2b); furthermore, when the lens touches the paper, object and image are almost the same size. (Try this yourself.) Since we do not use a magnifying glass in this manner, we do not realize that the object and image for a double convex lens can never be exactly equal in size. For the time being, we shall assume that this is possible, and later we will show the true behavior. The locations of object and image corresponding to unit magnification are called the *unit* or *principal points* H and H′, respectively, and the set of four points F, F′, H, and H′ are the *cardinal points*; the planes through these points normal to the axis of the lens are the *cardinal planes*.

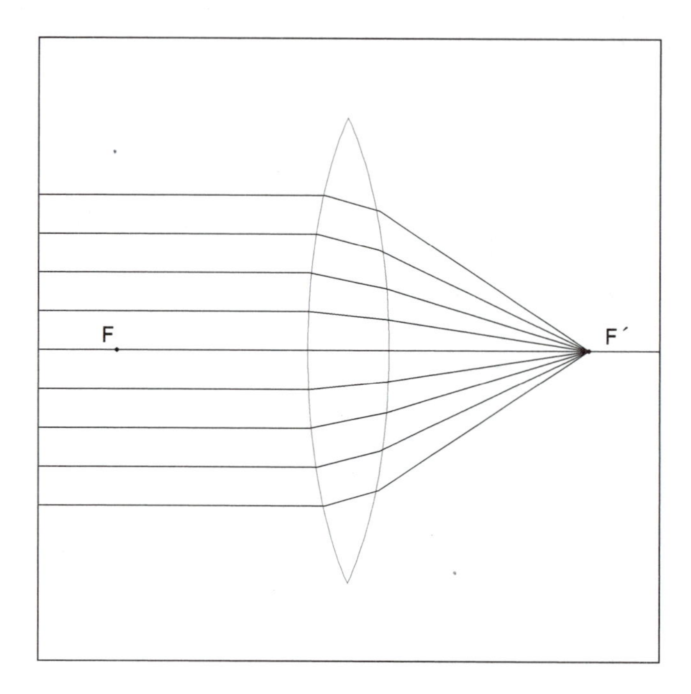

Fig. 1.1

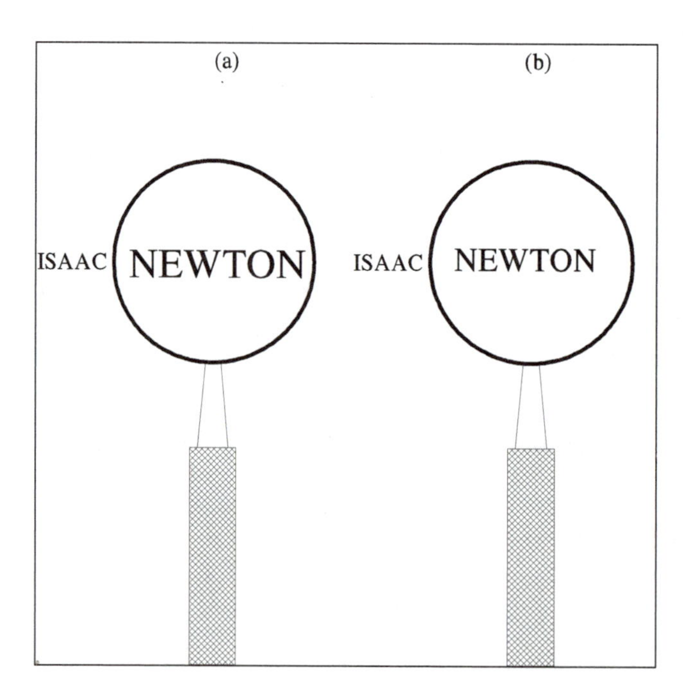

Fig. 1.2

We can now determine the image for the object in Figure 1.3. The coordinate orientation in this figure may appear strange at first. It is customary in optics to designate the lens axis as OZ, and the other two axes form a right-handed system. The x-axis then lies in the plane of the figure; the y-axis, coming out of the paper and perpendicular to it, is not shown. The object, of height x, is placed to the left of the focal point F. The cardinal points occur in the order F, H, H′, F′, since we expect—based on the experiment corresponding to Figure 1.2—that the unit points are very close to the lens. (Another fact to be demonstrated later is that H and H′ are actually *inside* the lens.) Two rays are shown leaving point P at the top of the object. The ray that is parallel to the axis strikes the unit plane through H and continues to the right, completely missing the lens. (This doesn't matter with the ray tracing procedure we are going to develop.) The ray then meets the unit plane through H′ at a point that *must* be a distance x from the axis. To understand why, we recognize that, if the object were relocated so as to lie on the object-space unit plane, the ray emerging at the other unit plane must be at a distance x from the axis, as required by the definition of these planes. We then assume that the ray is bent or *refracted* at the unit plane at H′ and proceeds to, and beyond, the focal point F′. (I will show, in Problem 10, that this requirement is consistent with the definition of a unit plane.)

The second ray, from P through the focal point F, is easily traced since the lens is indifferent to the direction of the light. Assume temporarily that we know the location of the image point P′. Let a parallel ray leave this point and travel to the left. The procedure just given indicates that this ray will strike the unit plane at H′, emerge at the other unit plane, pass through F, and reach the object point P. Then reversing the

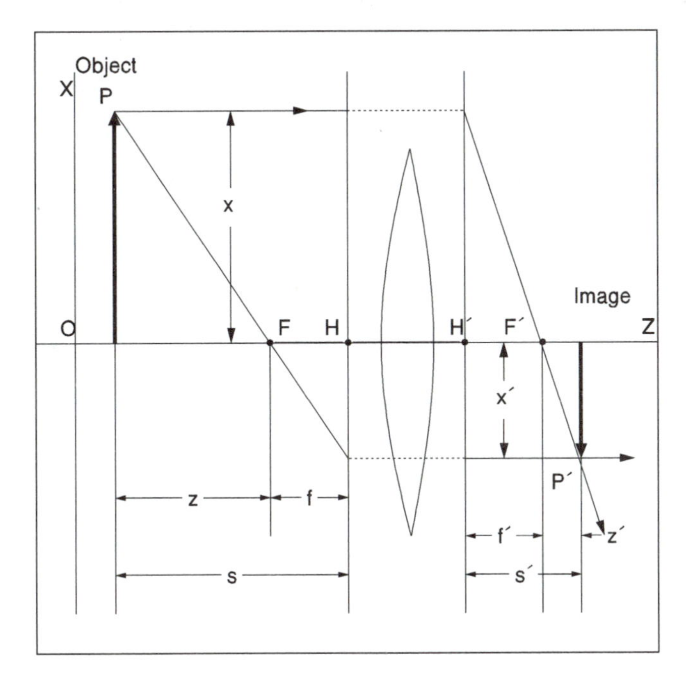

Fig. 1.3

direction of this ray, it will start at P, emerge at the unit plane H′, and travel parallel to the axis, intersecting the other ray at P′. Simple geometry gives the lens equations. The distance between F and H is called the *object space focal length* f, with f' being the corresponding image space quantity. These definitions may surprise you; physics books say that focal lengths should be measured from the center of the lens and we shall clarify this point a little later. There are three similar triangles on the object side of the lens, giving the relations

$$\frac{x + x'}{s} = \frac{x}{z} = \frac{x'}{f} \tag{1}$$

and, on the image side, in the same way

$$\frac{x + x'}{s'} = \frac{x'}{z'} = \frac{x}{f'} \tag{2}$$

where all quantities that appear in equations (1) and (2) are defined in Figure 1.3. It follows that

$$zz' = ff' \tag{3}$$

which is the *Newton lens equation*. If f and f' are the same, as we shall see later is usually the case, then (1) and (2) can be added to give

$$\frac{1}{s'} + \frac{1}{s} = \frac{1}{f} \tag{4}$$

This is the *Gauss lens equation* and it is equivalent to equation (3). Many authors advocate the use of the Newton equation because its structure is simpler. It turns out that the quantities z and z' are not usually known at the beginning of a problem, so that the Newton form of the lens equation is of little value. However, before we can use the Gauss equation, we need to establish sign conventions, and these will be introduced just a bit later.

PROBLEM 1

Prove that a ray in Figure 1.3 traveling from P to H and emerging at H′ will pass through P′, so that the rays PH and H′P′ are parallel.

Although it takes only two rays to determine the location of the image point in Figure 1.3, the third ray specified by Problem 1 serves as a convenient check on our ray tracing. (Add the rays PH and P′H′ to Figure 1.3 and confirm that they are parallel.) We note that the image in this figure is *real*, *inverted*, and *reduced*. That is, it is smaller than the object, oriented in the opposite direction, and can be projected onto a screen located at the image position. You can confirm this by using a magnifying glass to create an image of a distant light fixture using a sheet of white paper as a screen.

1.2 SNELL'S LAW

To consider how light rays behave in optical systems, we introduce the *index of refraction n* of a material medium, defined as the ratio of the velocity of light c in a vacuum to its velocity v in that medium, or

$$n = \frac{c}{v} \tag{5}$$

(The velocity of light in air is approximately the same as it is in a vacuum.) A light ray is shown in Figure 1.4 crossing the interface between air (the upper medium) and glass, with indices as labeled. Problem 2 will indicate a way of showing that the bending of the light at the interface is specified by the relationship

$$n \sin \theta = n' \sin \theta' \tag{6}$$

which is known as *Snell's law.*

PROBLEM 2

Fermat's principle states that a ray of light starting at some arbitrary point in one medium and crossing the interface of Figure 1.4 will arrive at an arbitrary point in the other medium in a time that is a minimum or a maximum. Use Figure 1.4 to show that Snell's law is a consequence of this condition.

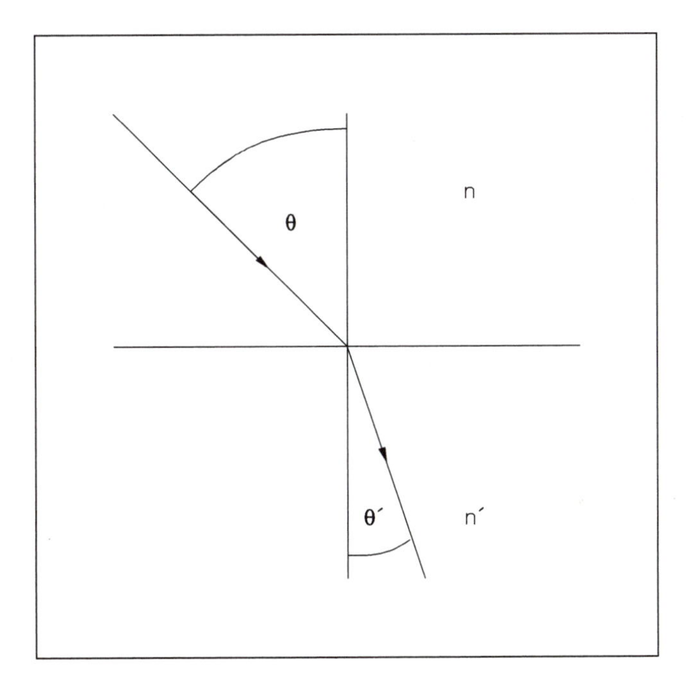

Fig. 1.4

1.3 PARAXIAL MATRICES

Figure 1.5 shows a ray of light leaving an object point P and striking the first surface of a lens at point P_1. It is refracted there and proceeds to the second surface. All rays shown lie in the z,x-plane, or *meridional plane*, which passes through the symmetry axis OZ. The amount of refraction is specified by Snell's law—equation (6)—which we shall now simplify. I remind you that the sine, cosine, and tangent of an angle can be expressed as a Taylor series. These have the form

$$\sin\theta = \theta - \frac{\theta^3}{3!} + \frac{\theta^5}{5!} - \dots \tag{7}$$

and

$$\cos\theta = 1 - \frac{\theta^2}{2!} + \frac{\theta^4}{4!} - \dots \tag{8}$$

where the angle θ is expressed in radians. For small values of θ, only the first term on the right of equation (7) or (8) need be used, as Problem 3 demonstrates.

PROBLEM 3

Show numerically that, for angles of 6° or less, the following approximations are valid.

$$\sin\theta = \theta, \quad \cos\theta = 1, \quad \tan\theta = \theta \tag{9}$$

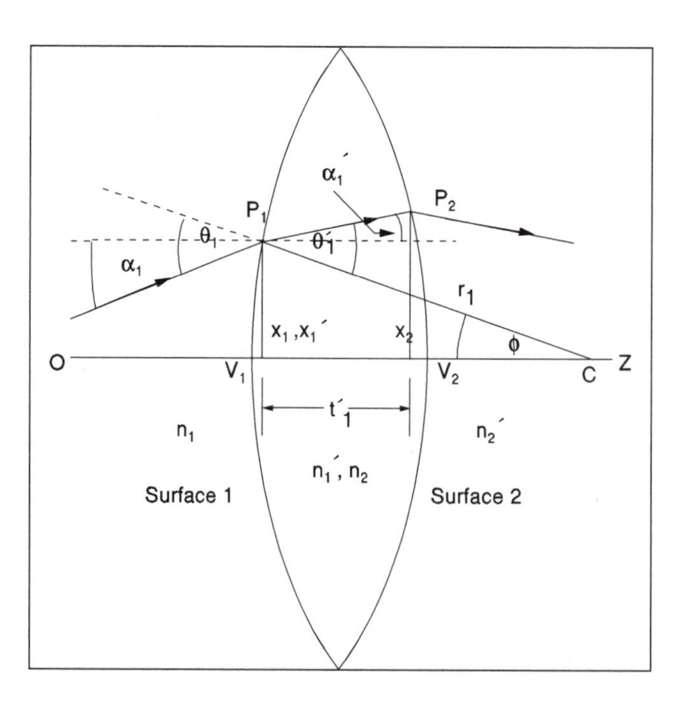

Fig. 1.5

If all the angles in Figure 1.5 are small, then Snell's law simplifies to

$$n_1\theta_1 = n'_1\theta'_1 \tag{10}$$

This is called the *paraxial* form of Snell's law; the word *paraxial* means "close to the axis." It turns out, however, that the Snell's law angles are not convenient to work with, and we eliminate them with the identities

$$\theta_1 = \alpha_1 + \phi, \quad \theta'_1 = \alpha'_1 + \phi \tag{11}$$

where α_1 is the angle that the incident ray makes with the axis OZ, α'_1 is the corresponding angle for the refracted ray, and ϕ is the angle that the radius r_1 (the line from C to P_1) of the lens surface makes at the center of curvature C. This particular angle can be specified as

$$\sin\phi = \frac{x_1}{r_1} \tag{12}$$

but we can use the paraxial approximation to simplify equation (12), obtaining

$$\phi = \frac{x_1}{r_1} \tag{13}$$

Substituting (11) and (13) into (10), the result can be written as

$$n'_1\alpha'_1 = \frac{n_1 - n'_1}{r_1}x_1 + n_1\alpha_1 \tag{14}$$

Notice that the distance from point P_1 to the axis is labeled as either x_1 or x'_1. This strange notation, which will be explained later, leads to the trivial relation

$$x_1 = x'_1 \tag{15}$$

which can be combined with (14) to obtain the matrix equation

$$\begin{pmatrix} n'_1\alpha'_1 \\ x'_1 \end{pmatrix} = \begin{pmatrix} 1 & -k_1 \\ 0 & 1 \end{pmatrix} \begin{pmatrix} n_1\alpha_1 \\ x_1 \end{pmatrix} \tag{16}$$

where the constant k_1 is called the *refracting power* of surface 1 and is defined as

$$k_1 = \frac{n'_1 - n_1}{r_1} \tag{17}$$

and where the first element $n'_1\alpha'_1$ in the one-column product matrix is calculated by multiplying the upper left-hand corner of the 2×2 matrix by the first element of the one-column matrix to its right, obtaining $1(n_1\alpha_1)$, and to this is added the product $-k_1x_1$

of the upper right-hand element in the 2 × 2 matrix and the second member of the one-column matrix. This result will bring back equation (14). A similar process for the second row of the 2 × 2 matrix gives equation (15). This square matrix is called the *refraction matrix* R_1 for surface 1, defined as

$$R_1 = \begin{pmatrix} 1 & -k_1 \\ 0 & 1 \end{pmatrix} \tag{18}$$

The *curvature* c_1 of lens surface 1 is defined as

$$c_1 = \frac{1}{r_1} \tag{19}$$

Then (17) may also be written as

$$k_1 = c_1(n'_1 - n_1) \tag{20}$$

Equation (20) indicates that the refracting power of a glass-air interface directly depends upon two parameters: the amount of curvature and the change in the index. Note that k_1 can be zero when the indices are equal and there is no refraction or when c_1 is zero (a flat glass plate) for which refraction occurs when the incident ray is not perpendicular to the plate.

Next we look at what happens to the ray as it travels from surface 1 to surface 2. As it goes from P_1 to P_2, its distance from the axis becomes

$$x_2 = x'_1 + t'_1 \tan \alpha'_1 \tag{21}$$

or, using the paraxial approximation for small angles

$$x_2 = x'_1 + t'_1 \alpha'_1 \tag{22}$$

Another approximation we shall use is to regard t'_1 as being equal to the distance between the lens vertices V_1 and V_2. Combining (22) with the identity

$$\alpha_2 = \alpha'_1 \tag{23}$$

leads to a second matrix equation

$$\begin{pmatrix} n_2\alpha_2 \\ x_2 \end{pmatrix} = \begin{pmatrix} 1 & 0 \\ t'_1/n'_1 & 1 \end{pmatrix} \begin{pmatrix} n'_1\alpha'_1 \\ x'_1 \end{pmatrix} \tag{24}$$

where the 2 × 2 *translation matrix* T_{21} is defined as

$$T_{21} = \begin{pmatrix} 1 & 0 \\ t'_1/n'_1 & 1 \end{pmatrix} \tag{25}$$

To get an equation that combines a refraction followed by a translation, we take the one-column matrix on the left side of (16) and substitute it into (24) to obtain

$$\begin{pmatrix} n_2\alpha_2 \\ x_2 \end{pmatrix} = T_{21}\ R_1 \begin{pmatrix} n_1\alpha_1 \\ x_1 \end{pmatrix} \tag{26}$$

Note that the 2×2 matrices appear in right-to-left order and that the multiplication of two square matrices is an extension of the rule given earlier. Specifically, to find the element in row i, column j, of the product, multiply the element in row i, column 1, of the left-hand matrix by the element in row 1, column j, of the right-hand matrix, then multiply the element in row i, column 2, of the left-hand matrix by the element in row 2, column j, of the right-hand matrix, and add these two products. Equation (26) gives the location and slope of the ray that strikes surface 2 after a refraction and a translation. To continue the ray trace at this surface, we introduce a second refraction matrix R_2 by expressing k_2 in terms of the two indices at this surface and the radius. Then equation (26) is extended to give the relation

$$\begin{pmatrix} n'_2\alpha'_2 \\ x'_2 \end{pmatrix} = R_2\ T_{21}\ R_1 \begin{pmatrix} n_1\alpha_1 \\ x_1 \end{pmatrix} \tag{27}$$

This process can obviously be applied to any number of refracting (and later to reflecting) surfaces; we are now in a position to trace rays within the limits of the paraxial approximation for a symmetrical optical system with an arbitrary number of components. The product of the three 2×2 matrices in equation (27) is known as the *system matrix* S_{21}, and it completely specifies the effect of the lens on the incident ray passing through it. It is also written as

$$S_{21} = R_2\ T_{21}\ R_1 = \begin{pmatrix} b & -a \\ -d & c \end{pmatrix} \tag{28}$$

where the four quantities a, b, c, and d are known as the *Gaussian constants*. They are used extensively in specifying the behavior of a lens or an optical system.

PROBLEM 4

Using the definitions (18) and (25), show that

$$a = k_1 + k_2 - \frac{k_1 k_2 t'_1}{n'_1} \tag{29}$$

$$b = 1 - \frac{k_2 t'_1}{n'_1} \tag{30}$$

$$c = 1 - \frac{k_1 t'_1}{n'_1} \tag{31}$$

$$d = \frac{-t'_1}{n'_1} \tag{32}$$

1.4 THE CONVENTIONS USED IN PARAXIAL MATRIX OPTICS

We now introduce sign and notation conventions, which are not only necessary for the application of paraxial matrix methods but will explain some of the things that previously seemed strange. The following rules are taken for the most part from E. L. O'Neill, *Introduction to Statistical Optics*, Addison-Wesley (1963).

1. Light normally travels from left to right.
2. The optical systems we deal with are symmetrical about the z-axis. The intersections of the refracting or reflecting surfaces with this axis are the vertices and are designated in the order encountered as V_1, V_2, and so on.
3. Positive directions along the axes are measured from the origin in the normal fashion, so that horizontal distances (that is, along the z-axis) are positive if measured from left to right. Angles are positive when measured up or out from the z-axis.
4. Quantities associated with the incident ray are unprimed; those for the refracted or reflected ray are primed.
5. A subscript denotes the associated surface.
6. If the center of curvature of a surface is to its right, the radius is positive, and vice versa.
7. A reflection reverses the sign of the index of refraction and of the translation direction.

This list seems formidable, but some of it should be familiar. A combination of rules 4 and 5 explains why the distance from P_1 to the axis in Figure 1.7 can have two labels; it specifies the end of the incident ray or the start of the refracted ray. Similarly, we see why the angle of inclination at P_2 can be designated as either α'_1, as appropriate for the refracted ray leaving surface 1, or α_2, which refers to the incident ray at surface 2. And the index of the glass could just as well be n'_1 or n_2. As we consider specific examples, you will find that these conventions are simple and logical. Rule 6 has the form shown because it will be found to agree with the other conventions, and rule 7 will be proven when we consider mirrors.

Example: A Double Convex Lens

In optics it is customary to work with a consistent set of units, such as inches or centimeters, and these need not be explicitly stipulated. A double convex lens has radii of 2.0 and 1.0, respectively; an index of refraction of 1.5; and a thickness of 0.5. These quantities are then denoted as

$$r_1 = +2.0,\ r_2 = -1.0,\ t'_1 = t_2 = +0.5$$

$$n_1 = 1.0,\ n'_1 = 1.5 = n_2,\ n'_2 = 1.0$$

From equation (17), we find that

$$k_1 = \frac{n'_1 - n_1}{r_1} = \frac{1.5 - 1.0}{2.0} = 0.25 \tag{33}$$

and similarly

$$k_2 = \frac{1.0 - 1.5}{-1.0} = 0.50 \tag{34}$$

We can find the Gaussian constants by using (29)–(32), or directly from the definition of the system matrix S_{21}, which is

$$S_{21} = R_2\, T_{21}\, R_1 = \begin{pmatrix} 1 & -0.50 \\ 0 & 1 \end{pmatrix} \begin{pmatrix} 1 & 0 \\ 0.5/1.5 & 1 \end{pmatrix} \begin{pmatrix} 1 & -0.25 \\ 0 & 1 \end{pmatrix} = \begin{pmatrix} 0.83 & -0.71 \\ 0.33 & 0.92 \end{pmatrix} \tag{35}$$

We note that the determinant of either a refraction matrix or a translation matrix has a value of unity, and hence the determinant of the system matrix should also have this value, since the determinant of the product of any number of matrices is the product of their determinants. This means that, for the matrix in (28), we have the relationship

$$bc - ad = 1 \tag{36}$$

We can verify this rule for the matrix in (35) to obtain

$$(0.83)(0.92) - (0.71)(-0.33) = 1.00$$

This property of the system matrix provides a very useful check on the accuracy of your matrix multiplications; use it regularly.

PROBLEM 5

A ray parallel to the axis strikes this lens at a distance of 2.0 units above the axis. Show that it emerges at a distance of 1.84 units above the axis and that it is pointing downward.

PROBLEM 6

A plano-convex lens has a first surface of radius equal to 10 units, the second surface is flat, and the thickness is 1 unit. The index is 1.5.

(a) Find the system matrix and check the values of the four elements. Verify that the refraction matrix for surface 2 is a unit matrix and has no effect on the system matrix.

(b) A ray strikes the lens at a distance of 2 units above the axis and at an inclination of +0.1. Find the position and angle of the emerging ray.

PROBLEM 7

A ray strikes a plane glass plate at an angle. Find the emerging position and angle using matrices.

1.5 Using the Gaussian Constants

Figure 1.6 shows the double convex lens of Figure 1.3 with the assumed locations of the cardinal points indicated. These positions, which we shall now determine accurately, are at distances designated as l_F, l_F', l_H, and l_H' and are measured from the associated vertex. The object position can be called t, a positive quantity that is measured to the right from the object to the first vertex, or it can be called t_1 if measured in the opposite direction. For the first choice, the matrix specifying the translation from object to lens will have the quantity t/n_1 in its lower left-hand corner. However, it is both logical and convenient to use the first vertex as the reference point, as we have already done for the focal and unit points. Hence, we replace t/n_1 in the translation matrix with the quantity $-t_1/n_1$ and remember to specify t_1 as a negative number when calculating the image position. The equation connecting object and image is obtained by starting with this matrix and multiplying it by the system matrix of equa-

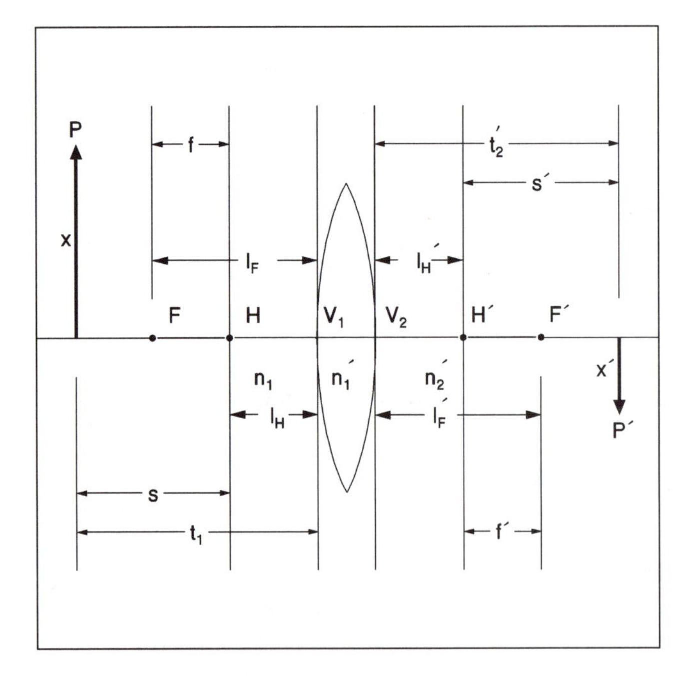

Fig. 1.6

tion (28), and finally by a translation matrix corresponding to the translation t_2' from lens to image to obtain

$$\begin{pmatrix} n'_2\alpha'_2 \\ x' \end{pmatrix} = \begin{pmatrix} 1 & 0 \\ t_2'/n'_2 & 1 \end{pmatrix} \begin{pmatrix} b & -a \\ -d & c \end{pmatrix} \begin{pmatrix} 1 & 0 \\ -t_1/n_1 & 1 \end{pmatrix} \begin{pmatrix} n_1\alpha_1 \\ x \end{pmatrix} \tag{37}$$

where α or α_1 is the angle that the ray from the top of the object makes with the z-axis. This equation takes the initial value of the inclination α and of the position x of the ray leaving the object and determines the final values α' and x' at the image. The product of the three 2×2 matrices on the right-hand side can be consolidated into a single matrix called the *object-image* matrix $S_{P'P}$. Then (37) can be written as

$$\begin{pmatrix} n_2'\alpha_2' \\ x' \end{pmatrix} = S_{P'P} \begin{pmatrix} n_1\alpha_1 \\ x \end{pmatrix} \tag{38}$$

Working out the algebra, the object-image matrix has the complicated form

$$S_{P'P} = \begin{pmatrix} b + \dfrac{at_1}{n_1} & -a \\ \dfrac{bt_2'}{n_2'} + \dfrac{at_1t_2'}{n_1n_2'} - d - \dfrac{ct_1}{n_1} & c - \dfrac{at_2'}{n_2'} \end{pmatrix} \tag{39}$$

PROBLEM 8

Verify (39) and show that the determinant of the matrix is unity.

If we put this matrix into (38), we obtain an unsatisfactory result: the value of x' will depend on the angle α_1 made by the incident ray, as can be seen by multiplying the two matrices on the right side. The ratio of x' to x is called the *magnification* m; that is

$$m = \frac{x'}{x} \tag{40}$$

A perfect image can be formed only if the magnification is determined solely by the object distance and the constants of the lens. To eliminate this difficulty, the lower left-hand element in the matrix (39) is required to be equal to zero, and the magnification is then

$$m = \frac{x'}{x} = c - \frac{at'_2}{n'_2} \tag{41}$$

Since the determinant of this matrix is unity, it also follows that

$$\frac{1}{m} = b + \frac{at_1}{n_1} \tag{42}$$

Equation (38) is now

$$\begin{pmatrix} n_2'\alpha_2' \\ x' \end{pmatrix} = \begin{pmatrix} 1/m & -a \\ 0 & m \end{pmatrix} \begin{pmatrix} n_1\alpha_1 \\ x \end{pmatrix} \tag{43}$$

PROBLEM 9

(a) Use the condition on the lower left-hand element of the matrix in (39) to show that

$$\frac{t_2'}{n_2'} = \frac{d + ct_1/n_1}{b + at_1/n_1}, \quad \frac{t_1}{n_1} = \frac{d - bt_2'/n_2'}{-c + at_2'/n_2'} \tag{44}$$

(b) Combine this with (41) to obtain (42).

This equation connects the object distance t_1 with the image distance t_2', both quantities being measured from the appropriate vertex. Previously, in the Gauss lens equation (4), we used s and s' to specify object and image distances with respect to H and H′, but equation (44) is more general and more useful.

We can determine the location of the unit planes by letting m = 1. Then t_2' in (41) is the location l'_H of the image-space unit plane, and it becomes

$$l'_H = \frac{n'_2(c - 1)}{a} \tag{45}$$

This relation expresses the location of the unit plane on the image side as the distance from V_2 to H′. Similarly, the location of H with respect to V_1 is

$$l_H = \frac{n_1(1 - b)}{a} \tag{46}$$

The system matrix of equation (35) indicates that both b and c are slightly less than 1, and a is positive. Then (46) gives a positive value for l_H, and (45) shows that l'_H is negative. This confirms the earlier statement that both H and H′ lie inside a double convex lens. To locate the focal planes, consider a set of parallel rays coming from infinity and producing a point image at F′. The terms containing t_1 in the left-hand part of (44) are much larger than d or b, and this relation becomes

$$l'_F = \frac{n'_2 c}{a} \tag{47}$$

Letting t_2' become infinite in the right-hand half of (44), the location of F is given by the equation

$$l_F = \frac{-n_1 b}{a} \tag{48}$$

The focal lengths f and f' shown in Figure 1.6 were defined earlier as the distances between their respective focal and unit planes. Hence

$$f' = l'_F - l'_H = \frac{n'_2}{a} \tag{49}$$

and

$$f = l_F - l_H = \frac{-n_1}{a} \tag{50}$$

If we let $n_1 = n'_2 = 1$ for air, these become

$$-f = f' = \frac{1}{a} \tag{51}$$

The Gaussian constant a is thus the reciprocal of the focal length in air. Even when the lens is not in air, it is still true that $f' = -f$ if the lens is in a single medium. It is also true regardless of whether or not the lens is symmetric. These last four equations further show that

$$l'_F = cf' \tag{52}$$

and

$$l_F = bf \tag{53}$$

Since $b = 0.83$ and $c = 0.92$ in (35), these two Gaussian constants give the fraction of f or f' that lies outside the lens. Figure 1.7 shows the cardinal planes and points (for clarity—the lens thickness is not to scale). We have thus confirmed what we said earlier about unit magnification: it cannot be achieved with a double convex lens, although we shall see later that it may be obtained approximately with a very thin lens.

PROBLEM 10

Let a ray, at 1 unit from the axis and parallel to it, pass through a double convex lens and be refracted at each surface. Show that, when this ray leaves surface 2, it coincides with a ray that would be refracted only once at H′, as we previously assumed.

PROBLEM 11

Confirm the locations of the cardinal points shown in Figure 1.7.

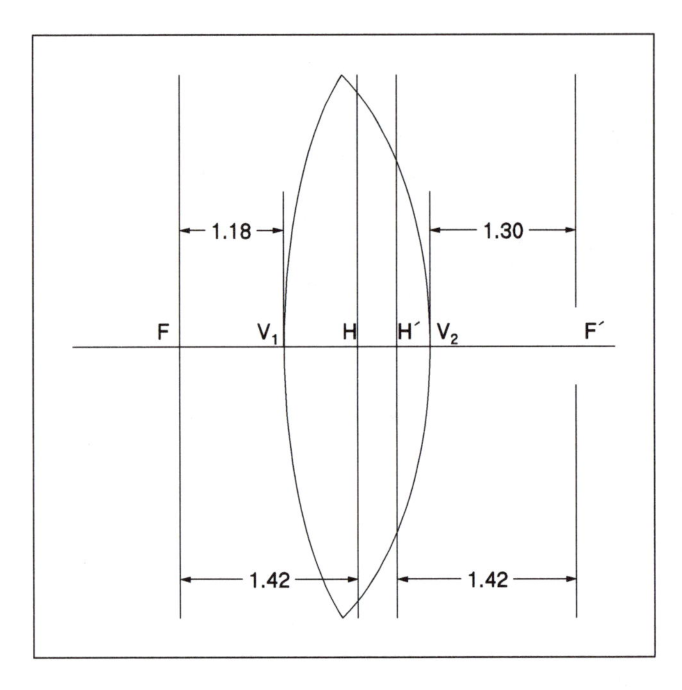

Fig. 1.7

PROBLEM 12

A glass sphere is used as a lens.

(a) Show that the unit points coincide at the center.

(b) Place an object of height 0.1 at V_1 for a sphere with a diameter of 15 units and an index of 1.5. Find the image both by calculation and by ray tracing.

PROBLEM 13

Place an object at vertex 1 of the lens in Figure 1.7. Locate the image in two ways and show that $m > 1$.

PROBLEM 14

Show that the focal lengths for a symmetrical lens in contact with two different media are unequal.

1.6 Ray Tracing

Since we know how to locate the cardinal points of a lens, we are in a position to ray trace in a precise manner, using the method of Figure 1.3 and adding a third ray as suggested by Problem 1. Once having located F, H, H′, and F′, we no longer need the lens to do the ray tracing; the cardinal planes with the proper separation are fully equivalent to the lens they replace. This is very much like what electrical engineers do when they replace a complicated circuit with a black box that has a pair of input and a pair of output terminals. This equivalent circuit behaves just like the original.

PROBLEM 15

Place an object of height $x = 2$ at a distance of 2.5 from the lens of Figure 1.7. Calculate the image size and position. Confirm your calculation by ray tracing to scale, using three sets of rays. (*Hint*: The sign of t_1 in [44] must be carefully chosen.)

Ray tracing becomes more complicated when we deal with *diverging* lenses (Figure 1.8). Parallel rays spread apart and do not come to a focus on the right-hand side of the lens. However, if the refracted rays are extended backwards, these extensions will meet as indicated. This behavior can be verified quantitatively by using the formulas developed earlier.

PROBLEM 16

A *double concave* lens is specified by the values $r_1 = -50$, $r_2 = +50$, $t'_1 = 15$, and $n'_1 = 1.5$. Find the cardinal points and the focal lengths.

You will find that F and F′ have exchanged the places that they have in a double convex lens, but H and H′ retain their original positions inside the lens. The ray extensions of Figure 1.8 are often necessary. If an object is placed between the focal point F

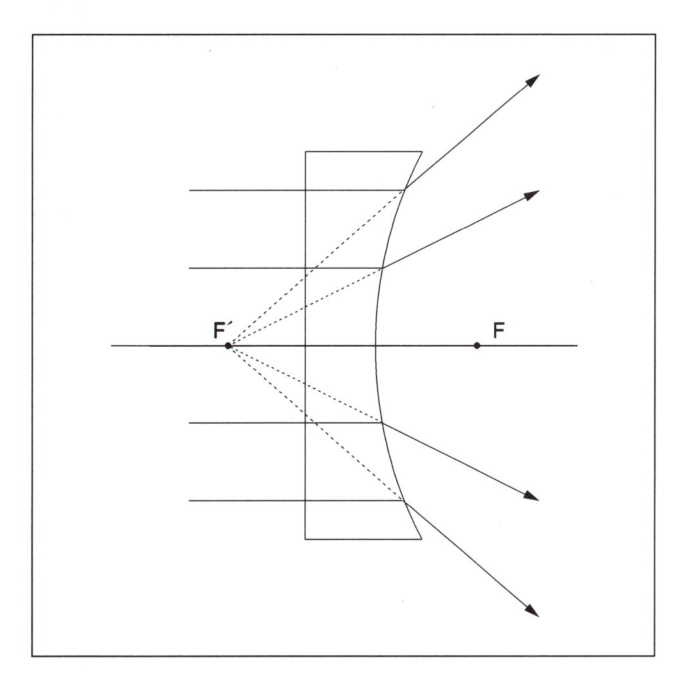

Fig. 1.8

and the first vertex of a convex lens (Figure 1.9), then our usual procedure produces two rays that do not intersect in image space. The ray from the object that is parallel to the axis is refracted downward so that it does not intersect the other ray in image space. However, the extensions to the left of these two rays meet to form an image that is *erect*, *magnified*, and *virtual*. This is the normal action of a magnifying lens; the image can be seen but cannot be projected onto a screen, unlike the real, inverted image of Figure 1.3. Note that this ray tracing diagram confirms what would be seen when a double convex lens is used as a magnifier.

PROBLEM 17

(a) Confirm Figure 1.9 by tracing the third set of rays.

(b) Trace the three sets of rays in Problem 13.

(c) Trace, to scale, an object at the unit plane.

PROBLEM 18

Given a lens for which $r_1 = 50$, $n'_1 = 1.5$, and $t'_1 = 15$, find r_2 such that an incident ray parallel to the axis remains parallel when it emerges. Such a lens is said to be *afocal*; its uses will be considered later.

For the double concave lens of Figure 1.10, a parallel ray leaves an object point P and is refracted upward at the unit plane H′ in such a way that its extension, rather

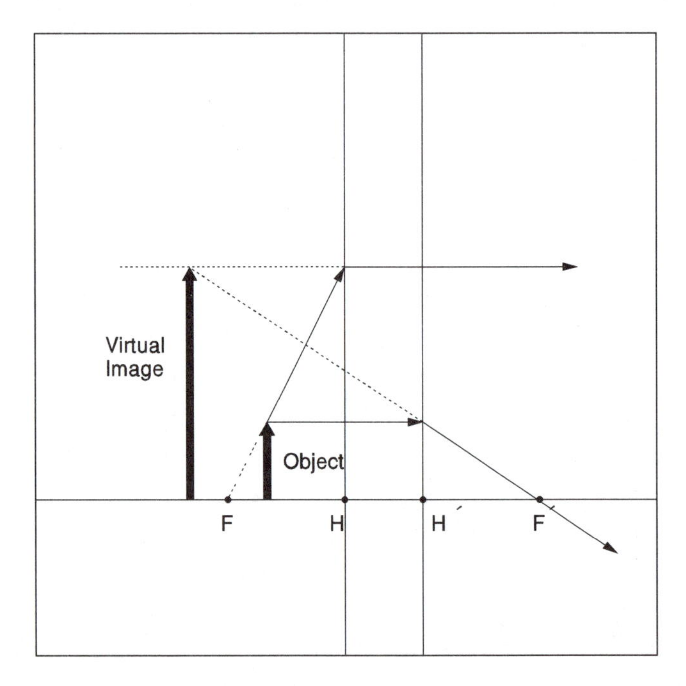

Fig. 1.9

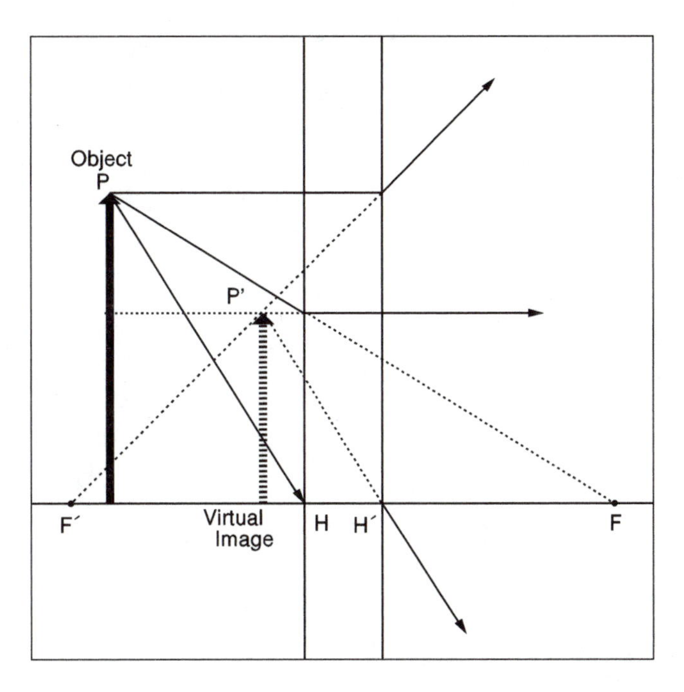

Fig. 1.10

than the ray itself, passes through F′. The ray headed for F is refracted at H before it can reach the object-space focal plane and becomes parallel to the axis. The third ray, going from P to H, emerges at H′ parallel to its original direction, and its extension to the left passes through the image point P′ already determined by the intersection of the other two rays. The resulting virtual image is upright and reduced. All three rays in this diagram behave exactly like the corresponding rays in Figure 1.3; the only change is the use of the extensions to locate the image. The eye receives the diverging rays from P and believes that they are coming from P′.

1.7 NODAL POINTS AND PLANES

Let a ray leave a point on the z-axis at an angle α_1 and have an angle α_2' when it reaches image space. Since $x = 0$ at the starting point, equation (43) shows that

$$n'_2\alpha'_2 = n_1\alpha_1/m \tag{54}$$

The ratio of the final angle to the initial angle in this equation is the *angular magnification* μ, which is then

$$\mu = n_1/mn'_2 \tag{55}$$

The locations of object and image for unit angular magnification are called the *nodal points*, labeled as N and N′. Equation (55) shows that the linear and angular magnification are reciprocals of one another when object and image are in air or in the same medium.

PROBLEM 19

Use (41) or (42) to show that the location of the nodal points with respect to the lens vertices is given by the relationships

$$\frac{l_N}{n_1} = \frac{(n'_2/n_1) - b}{a} \tag{56}$$

and

$$\frac{l'_N}{n'_2} = \frac{c - (n_1/n'_2)}{a} \tag{57}$$

PROBLEM 20

Show that the nodal points and the unit points are identical if the object and the image are in the same medium.

Problem 20 shows that, when an optical system is in air or a common medium, the nodal and unit points are identical, confirming the behavior of the rays from P to H and from H′ to P′ in Figures 1.3 and 1.10. They are parallel because they correspond to unit angular magnification.

1.8 Compound Lenses

A great advantage of the paraxial matrix method is the ease of dealing with systems having a large number of lenses. To start simply, consider the *cemented doublet* of Figure 1.11—a pair of lenses with surface 2 of the first element and surface 1 of the second element matching perfectly. The parameters of this doublet are given in the manner shown in Table 1.1. It is understood that the first entry under r is r_1, the second entry is r_2, and so on. Note that this doublet has only three surfaces. If the two parts were separated by an air space, producing an *air-spaced* doublet, then there would be four surfaces to specify. The values of the indices n' and the spacings t' are placed between the appropriate values of r.

Numbering the vertices in the usual manner, the system matrix is

$$S_{31} = R_3T_{32}R_2T_{21}R_1$$

where

$$k_2 = (n'_2 - n_2)/r_2 = (1.632 - 1.500)/{-2.0}$$

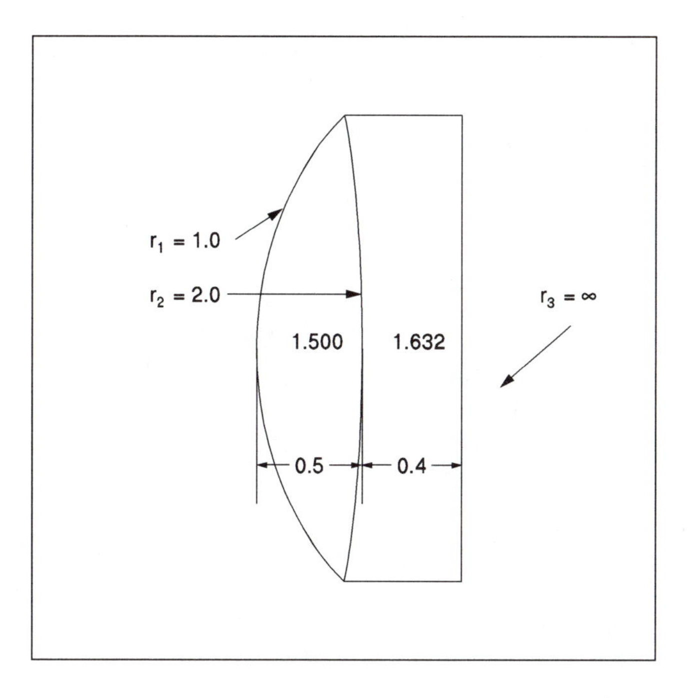

Fig. 1.11

Table 1.1 Doublet Specifications

r	*n*¢	*t*¢
1.0		
	1.500	0.5
–2.0		
	1.632	0.4
∞		

PROBLEM 21

Find the cardinal points and focal lengths of the doublet in Figure 1.11.

A lens used for many years in good cameras is the *Tessar*, which has three air-spaced components: a double convex lens, a double concave lens, and a cemented doublet. Its specifications are given in Table 1.2.

PROBLEM 22

Locating the cardinal points and planes for the Tessar involves the multiplication of a large number of matrices. This is a tedious process that can be made easier if done on a computer. Here is a program to accomplish this. The language used is QBASIC, which is included with the newer versions of Microsoft's DOS. The program starts by listing the radii, spacings, and indi-

Table 1.2 Tessar Specifications

r	*n¢*	*t¢*
1.628		
	1.6116	0.357
–27.57		
	1.0000	0.189
–3.457		
	1.6053	0.081
1.582		
	1.0000	0.325
∞		
	1.5123	0.217
1.920		
	1.6116	0.396
–2.400		

ces. Curvatures are used in the program to avoid infinite quantities. An object position of 10 units to the left of surface 1 is included, but this does not get used until later.

```
T(1) = 10: T(2) = .357: T(3) = .189: T(4) = .081
T(5) = .325: T(6) = .217: T(7) = .396
C(1) = 1 / 1.628: C(2) = -1 / 27.57: C(3) = -1 / 3.457
C(4) = 1 / 1.582: C(5) = 0: C(6) = 1 / 1.92: C(7) = -1 / 2.4
N(1) = 1: N(2) = 1.6116: N(3) = 1: N(4) = 1.6053
N(5) = 1: N(6) = 1.5123: N(7) = 1.61161
NP(1) = 1.6116: NP(2) = 1: NP(3) = 1.6053: NP(4) = 1
NP(5) = 1.5123: NP(6) = 1.6116: NP(7) = 1
```

The calculation starts by specifying the first refraction matrix R_1.

```
A11 = 1: A21 = 0: A12 = (N(1) - NP(1)) * C(1): A22 = 1
```

Then multiply this matrix by the product $T_{21}R_2$ of the first translation matrix and the second refraction matrix. This result in turn is multiplied by $T_{32}R_3$ and so on, using a BASIC FOR, NEXT loop.

```
FOR J = 2 TO 7
K(J) = (N(J) - NP(J)) * C(J)
B11(J) = 1 + K(J) * T(J) / N(J)
B12(J) = K(J)
B21(J) = T(J) / N(J)
B22(J) = 1
A3 = A11: A4 = A12
A11 = B11(J) * A11 + B12(J) * A21
A12 = B11(J) * A12 + B12(J) * A22
A21 = B21(J) * A3 + B22(J) * A21
A22 = B21(J) * A4 + B22(J) * A22
NEXT J
```

Note the use of temporary variables A3 and A4 to preserve the original values of A11 and A12. The Gaussian constants, the focal length, and the locations of the cardinal points and planes are then determined by the steps

```
A = -A12: B = A11: C = A22: D = -A21
FP = 1 / A: LFP = C / A: LF = -B / A: LHP = (C - 1) / A: LH = (1 - B) / A
```

To print

```
PRINT A, B, C, D
PRINT LH, LHP, LF, LFP, 1 / A12
```

(a) Use this program to find the locations of F, F′, H, and H′, as well as the values of f, f'.

(b) BASIC provides the following commands for achieving screen graphics.

`WINDOW(`x_1`, `y_1`) - (`x_2`, `y_2`)`: establishes the area in which graphics will appear by designating the coordinates of two diagonally opposite corners (x_1, y_1) and (x_2, y_2).

`LINE(`x_1`, `y_1`) - (`x_2`, `y_2`)`: draws a line between the two points (x_1, y_1) and (x_2, y_2).

`LOCATE(`x`, `y`)`: places text at the point (x, y).

Use these commands in conjunction with the program of (a) to produce the computer diagram shown in Figure 1.12.

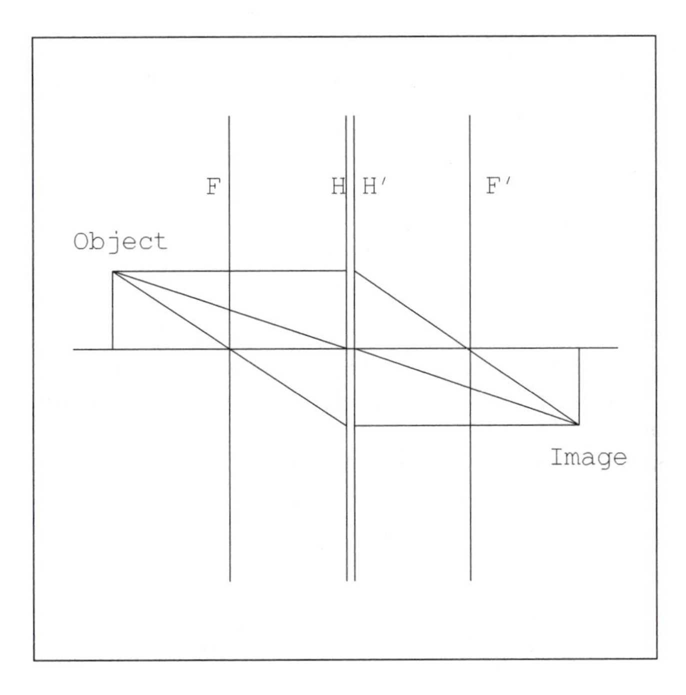

Fig. 1.12

1.9 APPROXIMATIONS FOR THIN LENSES

Among the nice features of paraxial matrix optics is the ease of obtaining useful relations. For example, if we combine the several expressions we have for the Gaussian constant a, we obtain

$$a = -\frac{n_1}{f} = \frac{n'_2}{f'} = k_1 + k_2 - \frac{k_1 k_2 t'_1}{n'_1} \tag{58}$$

Using the definition (17) with the lens in air, this leads to

$$\frac{1}{f'} = (n'_1 - 1)\left[\frac{1}{r_1} - \frac{1}{r_2} + \frac{(n'_1 - 1)t'_1}{n'_1 r_1 r_2}\right] \tag{59}$$

which is the *lensmakers' equation*. It tells how to find the focal length of a lens from a knowledge of its material and geometry. This derivation is much more direct than what you will usually find. Some optics books find the image produced by the first surface of a lens and use it as a *virtual object* for the second surface. This leads to confusion about signs and to messy algebra as well. The multiplication of matrices is more direct and easier to understand. This equation becomes simpler when we are dealing with *thin lenses*—those for which the third term in brackets can be neglected by assuming that the lens thickness is approximately zero. Then

$$\frac{1}{f'} = (n'_1 - 1)\left[\frac{1}{r_1} - \frac{1}{r_2}\right] \tag{60}$$

which is the form often seen in texts. We then realize that (59) is valid for lenses of any thickness, but (60) can be used if the spacing is much less than the two radii. We can also see that (30) and (31) reduce to

$$b = c = 1 \tag{61}$$

and then (45) and (46) become

$$l_H = l'_H = 0 \tag{62}$$

Thus, the unit planes are now at the vertices, and an object at V_1 would produce an image of the same size and orientation at V_2. Since H and H′ are now very close, they can be superimposed for convenience. A ray from any point on the object would pass undeviated through the geometrical center of the lens, as is often shown in elementary optics books. Equation (60) has an interesting interpretation; it can be written as

$$a = (n'_1 - 1)(c_1 - c_2) \tag{63}$$

indicating that the Gaussian constant a is proportional to both the change in index and the difference in curvatures between the two surfaces. When the thickness is negligible, we can let $d = 0$ and these new values of the Gaussian constants give a system matrix of the form

$$S_{21} = \begin{pmatrix} 1 & -1/f' \\ 0 & 1 \end{pmatrix} \tag{64}$$

This matrix contains a single constant—the focal length of the lens—so that the preliminary design of an optical system is merely a matter of specifying the focal length of the lenses involved and using the resulting matrices to determine the object-image relationship. We also note that this matrix looks like a refraction matrix; the nonzero element is in the upper right-hand corner. This implies that the refraction-translation-refraction procedure that actually occurs can be replaced, for a thin lens, by a single refraction occurring at the coinciding unit planes. Opthalmologists take advantage of the thin lens approximation by specifying focal lengths in units called *diopters*. The reciprocal of the focal length in meters (m) determines its refraction power in diopters. For example, a lens with $f' = 10$ centimeters (cm) will be a 10-diopter lens. If two lenses are placed in contact, the combined power is the sum of the individual powers, for multiplying matrices like equation (64) gives an upper right-hand element of the form $(-1/f'_1 - 1/f'_2)$.

1.10 THIN LENS COMBINATIONS

Consider the *compound magnifier* of Figure 1.13, which has an object 60 units to its left. The system matrix is found by multiplying the converging lens matrix, the translation matrix for a separation of 12 units, and the diverging lens matrix to obtain

$$S_{41} = \begin{pmatrix} 1 & 1/10 \\ 0 & 1 \end{pmatrix} \begin{pmatrix} 1 & 0 \\ 12 & 1 \end{pmatrix} \begin{pmatrix} 1 & -1/15 \\ 0 & 1 \end{pmatrix} \tag{65}$$

or

$$S_{41} = \begin{pmatrix} 11/5 & -7/150 \\ 12 & 1/5 \end{pmatrix} \tag{66}$$

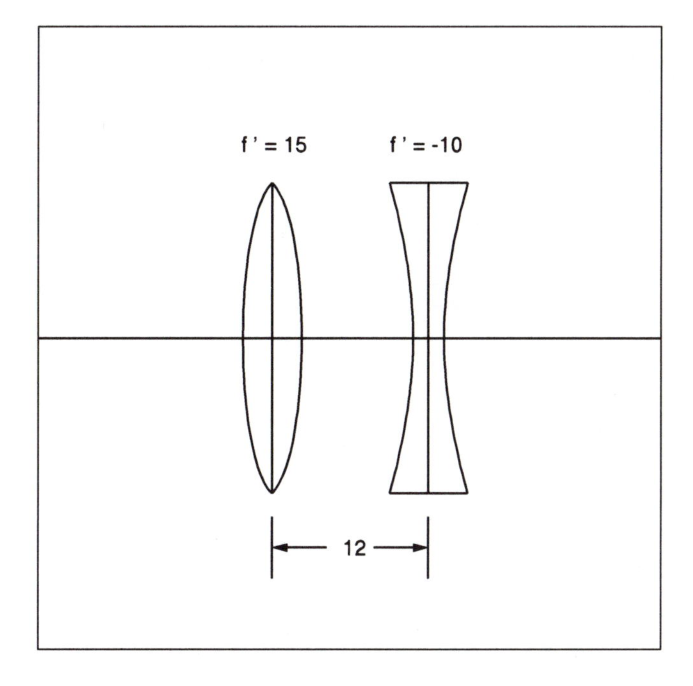

Fig. 1.13

The Gaussian constants are

$$a = 7/150,\ b = 11/5,\ c = 1/5,\ d = -12$$

with

$$bc - ad = 11/25 + 84/150 = 1$$

The image position is

$$t' = \frac{(1/5)(-60) - 12}{(7/150)(-60) + (11/5)} = 40 \tag{67}$$

and the magnification, computed in two ways as a check, is

$$\frac{1}{m} = b + at' = -3/5 \tag{68}$$

or

$$m = c - at = (1/5) - (-7/150)(-40) = -5/3 \tag{69}$$

That is, the image is somewhat larger than the object and inverted. The locations of the cardinal planes are

$$l_H = -25.71,\ l'_H = -17.14$$

$$l_F = -47.14,\ l'_F = 4.29$$

using equations (45)–(48). These distances are shown approximately to scale in Figure 1.14. The ray tracing then generates an enlarged, inverted real image, verifying the calculations.

Rays can also be traced numerically; for example, one starting at 10 units above the axis and parallel to it will arrive at the image with position and angle as determined by the equation

$$\begin{pmatrix} \alpha'_2 \\ x' \end{pmatrix} = \begin{pmatrix} -3/5 & -7/150 \\ 0 & -5/3 \end{pmatrix} \begin{pmatrix} 0 \\ 10.0 \end{pmatrix} = \begin{pmatrix} -0.47 \\ -16.7 \end{pmatrix} \tag{70}$$

so that

$$\alpha'_2 = -0.47,\ x' = -16.7$$

PROBLEM 23

(a) Find the system matrix for two thin lenses of focal length f'_1 and f'_2, respectively, and show that

$$a = \frac{1}{f'} = \frac{1}{f'_1} + \frac{1}{f'_2} - \frac{t'_2}{f'_1 f'_2},\quad b = 1 - \frac{t_2'}{f_2'},\quad c = 1 - \frac{t_2'}{f_1'},\quad d = -t_2' \tag{71}$$

(b) Verify the focal length formula for the lenses of Figure 1.14.

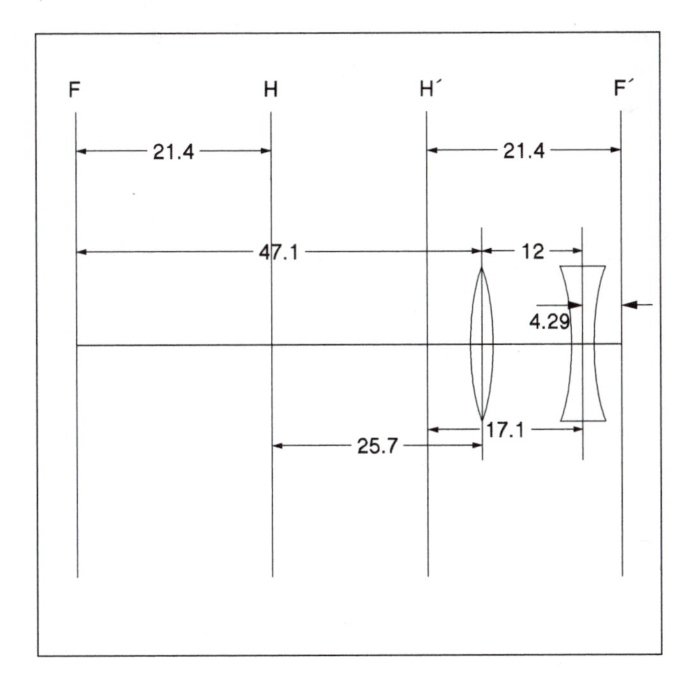

Fig. 1.14

PROBLEM 24

(a) Show by multiplying the exact rather than the approximate matrices that the focal length of a pair of thick lenses is given by the formula

$$\frac{1}{f'} = \frac{1 - k_4 t'_3 / n'_3}{f'_1} + \frac{1 - k_1 t'_1 / n'_1}{f'_2} - \frac{t'_2}{f'_1 f'_2} \tag{72}$$

(b) Show that this reduces to (71) for thin lenses.

PROBLEM 25

(a) A pair of thin lenses with focal lengths $f'_1 = 4$ and $f'_2 = 6$ and a spacing of 2 units has an object of arbitrary height at a distance of 10 units from the first vertex. Find the image by tracing three rays.

(b) Repeat for focal lengths 2 and 5, a spacing of 10, and an object at 3 units from the vertex.

PROBLEM 26

(a) Show that the object-image matrix for any lens is

$$S_{P'P} = \begin{pmatrix} 1 + as/n_1 & -a \\ 0 & 1 - as'/n'_2 \end{pmatrix} \tag{73}$$

where the locations of the object and the image are given by s and s', as measured from the unit planes in the appropriate direction in Figure 1.6.

(b) From this, show that the Gauss lens equation is

$$\frac{n'_2}{f'} = -\frac{n_1}{f} = -\frac{n_1}{s} + \frac{n'_2}{s'} \tag{74}$$

For a thin lens in air, (74) is like (4), but it incorporates the correct sign for s.

(c) Show that

$$m = s'n_1/sn'_2 \tag{75}$$

(d) Show from (44) that for a thin lens

$$\frac{n'_2}{f'} = \frac{-n_1}{t_1} + \frac{n_2}{t'_2} \tag{76}$$

and that (74) is identical to (76). Note that (44) is valid for any lens, thin or thick, and hence can be called the *generalized Gauss lens equation.*

PROBLEM 27

(a) Let D be the total distance between an object and its image as formed by a thin lens. Show that

$$D = [2 - \frac{1}{m} - m]f' \tag{77}$$

(b) Show that

$$f' = \frac{-mD}{(m-1)^2}, \quad s' = ms = \frac{mD}{m-1} \tag{78}$$

Also show that

$$s = \frac{D}{m-1} \tag{79}$$

These relations are very useful. If you know how far apart you want the object and image to be and what magnification you need, you can determine the focal length and location of the appropriate lens.

PROBLEM 28

(a) Determine f' and s for an object-image distance of 31.25 and a magnification of $-1/4$.

(b) Show that (74) in air gives relationships of the form

$$f' = \frac{ms}{m-1} = -\frac{ss'}{D} \tag{80}$$

PROBLEM 29

Make a scale drawing, trace two rays, and calculate s' for the following situations:

(a) Converging thin lens, focal length of 2 units, object at 4 units.

(b) Converging thin lens, focal length of 4 units, object at 2 units.

(c) Diverging thin lens, focal length of 4 units, object at 2 units.

PROBLEM 30

A thin lens of index 1.52 has a focal length of 10 units in air and 50 units in a liquid. Find the index of the liquid.

PROBLEM 31

Four thin lenses of focal lengths –1, 4/3, –1/3, and 1 have spacings of 1, 3, and 1/2, respectively. Show that this combination is equivalent to no lens at all in either direction.

Let us now consider the application mentioned in Problem 27 by specifying a spacing D between object and image and a desired magnification m, and then determining the focal length of the thin lens that will meet these requirements. We also want the location of object and image with respect to the lens. Using the formulas of Problems 26–28, we start with (75). In air

$$m = s'/s,\ s' = ms$$

Then

$$D = -s + s' = s(m-1)$$

or

$$s = D/(m-1)$$

and

$$s' = ms = mD/(m-1)$$

By (74)

$$f' = -ss'/D$$

As an example, let object and image be 62.5 units apart, and let the real, inverted image be one-fourth the size of the object. Then

$$s = 62.5/[-0.25 - 1] = -50.0$$

$$s' = (-1/4)(-50.0) = 12.5$$

$$f' = -(12.5)(-50.0)/62.5 = 10.0$$

This may be confirmed with (77), which gives

$$D = [2 + 4 + 0.25]10.0 = 62.5$$

1.11 REFLECTORS

To show how reflecting surfaces are handled with matrices, consider a plane mirror located at the origin and normal to the z-axis (Figure 1.15). The ray, which strikes it at an angle α_1, leaves at an equal angle, resulting in *specular reflection*. Since k = 0 for a flat surface, the refraction matrix reduces to the unit matrix and by equation (16)

$$n'_1\alpha'_1 = n_1\alpha_1$$

This contradicts Figure 1.15, since α_1 and α'_1 are opposite in sign. If, however, we specify that

$$n'_1 = -n_1$$

then the difficulty is removed and we have confirmed rule 7 in the listing of sign conventions.

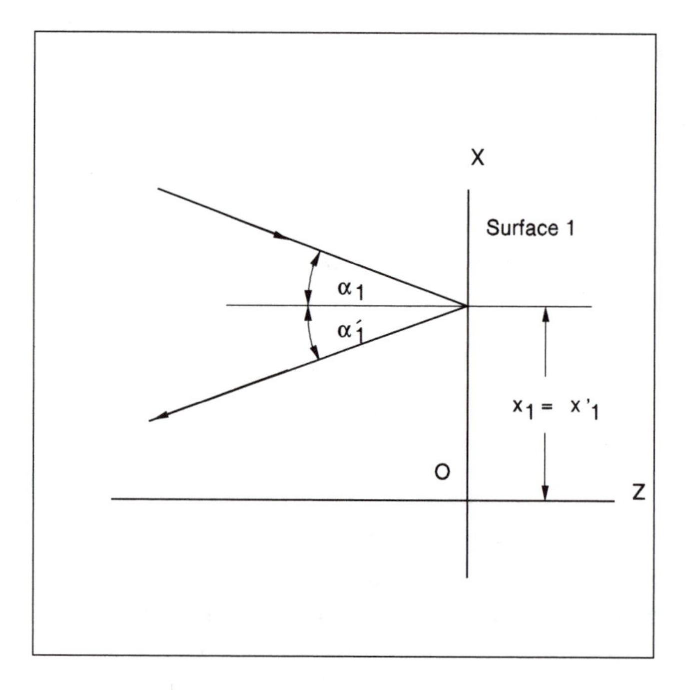

Fig. 1.15

PROBLEM 32

(a) Write the system matrix for a plane mirror and use it to show that F and F′ are infinitely far from the vertex.

(b) Write the object-image matrix for a plane mirror and use it to locate the image. Since the focal points cannot be used, confirm your result by tracing real rays.

The way rule 7 works for successive reflections can be explained with the two parallel mirrors of Figure 1.16. A ray making a positive angle α_1 strikes the first mirror and is reflected with a negative slope. The translation to the second mirror is governed by the equation

$$\begin{pmatrix} n_2\alpha_2 \\ x_2 \end{pmatrix} = \begin{pmatrix} 1 & 0 \\ t'_1/n'_1 & 1 \end{pmatrix} \begin{pmatrix} n'_1\alpha'_1 \\ 0 \end{pmatrix} \tag{81}$$

or

$$n_2\alpha_2 = n'_1\alpha'_1, \quad x_2 = \alpha'_1 t'_1 \tag{82}$$

Since α'_1 (which is the same as α_2) and t'_1 are both negative, then x_2 is positive, as shown. For the second reflection, using $n'_2 = -n_2$ in the reflection equation, we obtain

$$\alpha'_2 = -\alpha_2, \quad x'_2 = x_2 \tag{83}$$

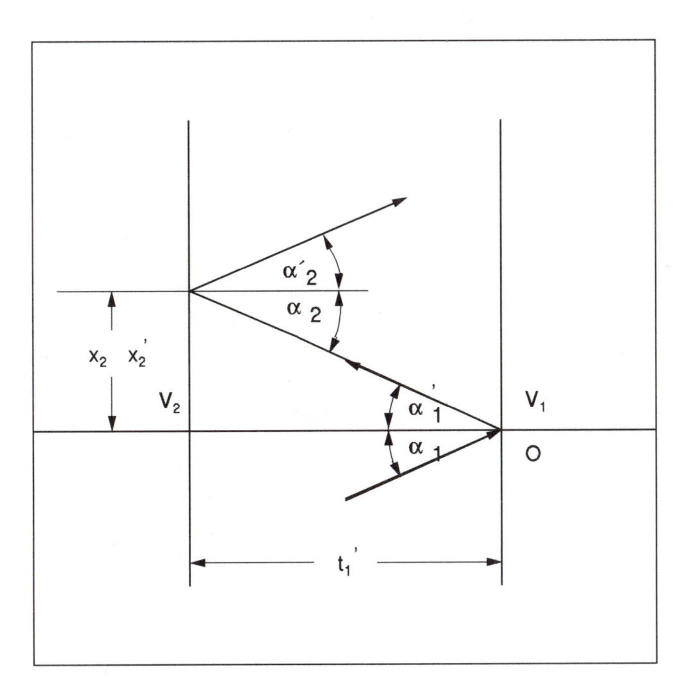

Fig. 1.16

and α'_2 is positive since α_2 is negative. Thus, the second reversal of sign by the index gives the expected results.

1.12 Spherical Mirrors

The image created by a concave spherical mirror can be found without the use of the cardinal points and planes. Figure 1.17 shows a ray leaving the object point P, traveling parallel to the axis until it is reflected at A, making equal incident and reflected angles with respect to the radial line from the center C. The ray striking the vertex V_1 behaves in the same way, and a ray through C will come back on itself. For small angles, all three rays meet at P′. To apply paraxial matrix methods to this reflector, the refraction power is

$$k_1 = \frac{n'_1 = n_1}{r_1} = \frac{-1 - (1)}{r_1} = -\frac{2}{r_1} \tag{84}$$

The system matrix for a single refracting surface thus becomes

$$S_{11} = R_1 = \begin{pmatrix} 1 & -k_1 \\ 0 & 1 \end{pmatrix} = \begin{pmatrix} 1 & 2/r_1 \\ 0 & 1 \end{pmatrix} \tag{85}$$

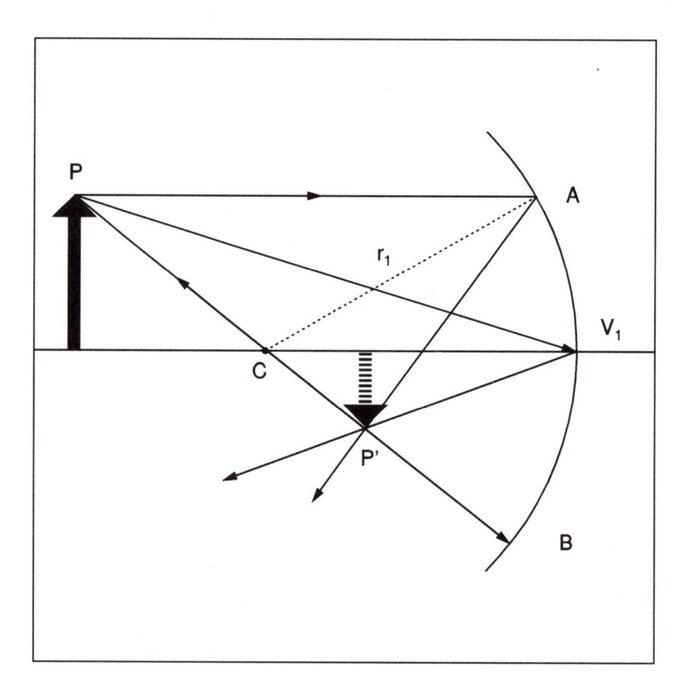

Fig. 1.17

$$a = -2/r_1, \quad b = c = 1, \quad d = 0 \tag{86}$$

The connection between object and image can be expressed as

$$\begin{pmatrix} n'_1\alpha'_1 \\ x'_1 \end{pmatrix} = \begin{pmatrix} 1 & 0 \\ t'/n'_1 & 1 \end{pmatrix}\begin{pmatrix} b & -a \\ -d & c \end{pmatrix}\begin{pmatrix} 1 & 0 \\ -t/n_1 & 1 \end{pmatrix}\begin{pmatrix} n_1\alpha_1 \\ x_1 \end{pmatrix} \tag{87}$$

and, using the same procedure that was applied to equation (37), the locations of the six cardinal points are

$$l_F = -n_1 b/a, \quad l'_F = n'_1 c/a \tag{88}$$

$$l_H = n_1(1 - b)/a, \quad l'_H = n'_1(c - 1)/a \tag{89}$$

$$\frac{l_N}{n_1} = \frac{(n'_1/n_1) - b}{a}, \quad \frac{l'_N}{n'_1} = \frac{c - (n_1/n'_1)}{a} \tag{90}$$

so that for the spherical mirror

$$l_F = \frac{(-1)(1)}{-2/r_1} = \frac{r_1}{2} = l'_F \tag{91}$$

$$l_H = l'_H = 0 \tag{92}$$

and

$$l_N = \frac{-1 - 1}{-2/r_1} = r_1 \tag{93}$$

but

$$\frac{l'_N}{-1} = \frac{1 - (-1)}{-2/r_1}, \quad l'_N = r_1 \tag{94}$$

Since r_1 is negative, these equations show that the unit points lie on the vertex, the foci coincide halfway between the center of curvature and the vertex, and the nodal points are at this center. The perfect focusing is a consequence of the paraxial approximation; the true situation will be considered later. Ray tracing for this mirror uses the procedure developed for lenses, which applies to optical systems of any complexity. Figure 1.18 shows an object to the left of the center of curvature. The ray from P parallel to the axis goes to H′ and then through F′, while the ray through F goes to H and then becomes parallel to the axis. Their intersection at P′ produces a real, inverted image. The third set of rays represents something different—it requires a knowledge of the nodal point locations. The ray from P to N should be parallel to the ray from N′ to P′, by definition; in this example, they meet this requirement by being colinear.

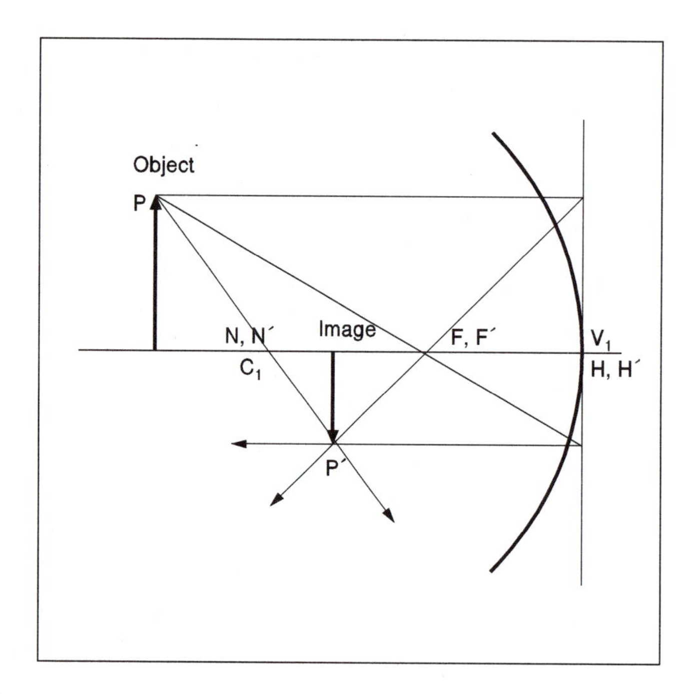

Fig. 1.18

PROBLEM 33

Find the radius of a makeup/shaving mirror that gives an erect image of twice the size of the viewer, who is 25 cm away. Do the ray tracing.

PROBLEM 34

An object with height 15 units is 60 units from a convex mirror of radius 0.8. Describe the image numerically and graphically.

PROBLEM 35

A mirror lies 20 units below the surface of a bowl of water. A lens, $f' = 2$, sits on the mirror. Light rays leave a point P above the surface and are reflected back on themselves. The index of the lens is 1.5, and for water it is 4/3. Where is P located?

PROBLEM 36

An object is 15 units to the left of a converging lens, $f' = 10$. A concave mirror, $r = 16$, is 20 units to the right of the lens. Determine the cardinal points and find the image by ray tracing if

(a) The rays pass through the lens only once.

(b) The rays pass through the lens again after reflection from the mirror.

(*Hint*: Be very careful about the index after reflection.)

PROBLEM 37

Figure 1.19 shows a *Cassegrain* telescope. The large concave *primary* mirror has an aperture at its center to permit the rays from the smaller convex *secondary* mirror to reach an eyepiece or a photographic plate. Let the concave mirror have a radius of 200, the convex mirror have a radius of 50, and the separation be 80 units. Locate the cardinal points, find the focal length, and determine the image of a distant star.

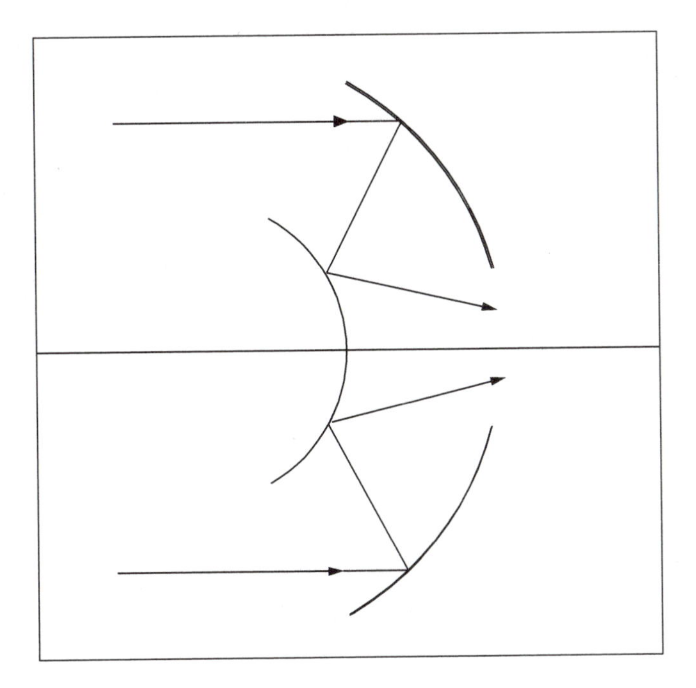

Fig. 1.19

1.13 RETROREFLECTORS

The *retroreflector* or *cat's eye* reflects a ray of light in a direction parallel to the incident ray. One of the main uses is in road safety markers. A simple way to make a retroreflector is to coat one side of a transparent sphere with a reflecting film. A ray striking the uncoated side will behave as shown in Figure 1.20. The matrix equation governing the refraction at the first vertex and the translation to the second vertex (dropping subscripts for convenience) is

$$\begin{pmatrix} n'\alpha' \\ x' \end{pmatrix} = \begin{pmatrix} 1 & 0 \\ \dfrac{2r}{n'} & 1 \end{pmatrix} \begin{pmatrix} 1 & \dfrac{-(n'-1)}{r} \\ 0 & 1 \end{pmatrix} \begin{pmatrix} n\alpha \\ x \end{pmatrix} = \begin{pmatrix} 1 & \dfrac{1-n'}{r} \\ \dfrac{2r}{n'} & \dfrac{2}{n'} - 1 \end{pmatrix} \begin{pmatrix} n\alpha \\ x \end{pmatrix} \tag{95}$$

Since the initial slope is zero ($\alpha = 0$) and we want the position at V_2 to be zero ($x' = 0$), these values in (95) give

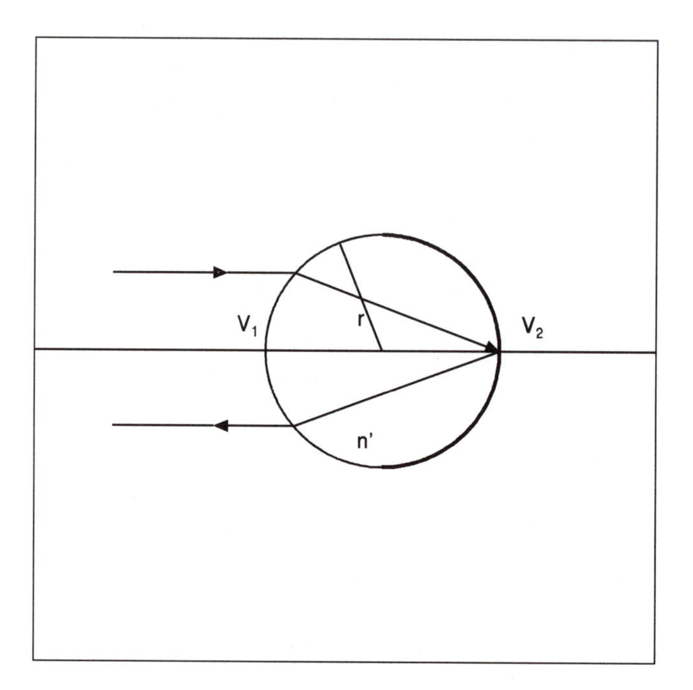

Fig. 1.20

$$n' = 2 \tag{96}$$

Although an index as high as this is not possible for glass, beads meeting this criterion can be made from plastic and are embedded in road signs.

PROBLEM 38

(a) Label the vertices of the sphere as V_1, V_2, and V'_1 to denote that the ray returns to the first surface after reflection. Taking into account the two refractions, the reflection, and a translation in each direction, show that the system matrix will be

$$S_{1'1} = \begin{pmatrix} \dfrac{n' - 4}{n'} & \dfrac{2(n' - 2)}{n'r} \\ -\dfrac{4r}{n'} & \dfrac{n' - 4}{n'} \end{pmatrix} . \tag{97}$$

where r is the absolute value of the radius. Be careful about signs for n' after refraction and for r at both surfaces.

(b) Use this matrix to show that $n' = 2$ for retroreflector behavior.

A simpler retroreflector is the plane and spherical mirror combination of Figure 1.21. A parallel incoming ray is reflected as shown in the figure; this behavior can be verified by a matrix equation.

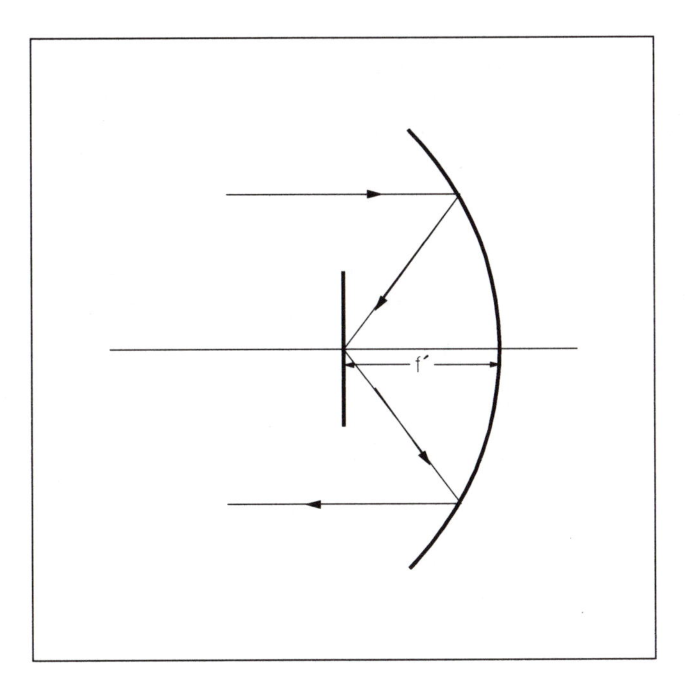

Fig. 1.21

PROBLEM 39

(a) Show that the system matrix for this arrangement is

$$S_{31} = \begin{pmatrix} -1 & 0 \\ r & -1 \end{pmatrix} \tag{98}$$

(b) Use it to confirm the behavior shown in Figure 1.21.

(c) Let a ray strike the concave mirror at an arbitrary angle α'. Show that it emerges parallel to itself.

(d) Confirm this with a scale diagram.

PROBLEM 40

A thin lens of focal length f' is in front of a concave mirror with radius equal to f' and its center of curvature at the lens. Show that this system is a retroreflector with system matrix $-I$, where I is the unit matrix, defined as

$$I = \begin{pmatrix} 1 & 0 \\ 0 & 1 \end{pmatrix} \tag{99}$$

PROBLEM 41

A plano-convex lens and a mirror (Figure 1.22) are arranged so that the centers of curvature of both the mirror and the lens are at V_1 and the radii are related such that

$$r_L = \frac{(n - 1)r_M}{n} \tag{100}$$

where r_L is the lens radius, r_M is the mirror radius, and the two radii are expressed as absolute values. Show that the system matrix for this retroreflector is

$$S_{1'1} = \begin{pmatrix} -1 & 0 \\ 0 & -1 \end{pmatrix} \tag{101}$$

Since surface 1 is flat, start with the translation from surface 1 to surface 2. Be sure to use the negative signs for the radii.

The form of the system matrix for each of the four retroreflectors we have considered is interesting. If we substitute $n' = 2$ into (97), we obtain

$$S_{1'1} = \begin{pmatrix} -1 & 0 \\ -2r & -1 \end{pmatrix} \tag{102}$$

This looks like a translation with the signs reversed on the diagonal and can be regarded as the product of a normal translation matrix and the matrix $-I$. In Figure

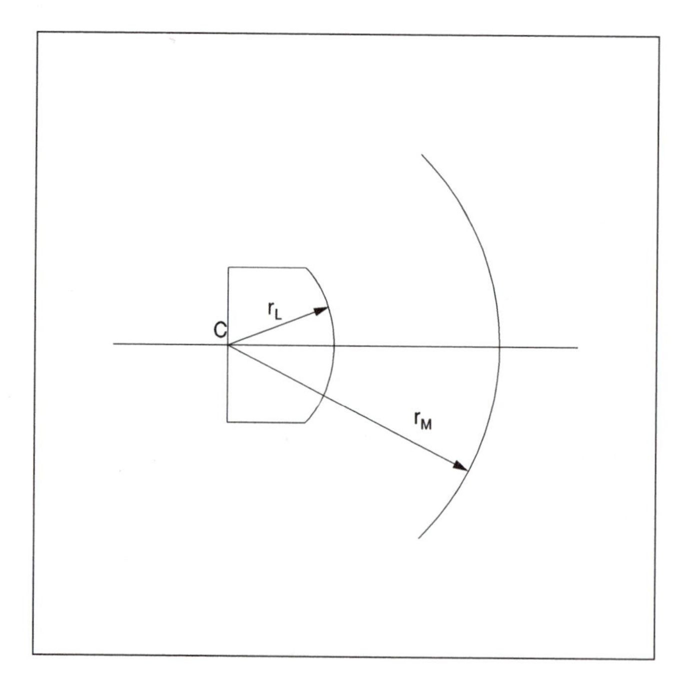

Fig. 1.22

1.20, there are two translations of magnitude $2r/n'$, where both r and n' reverse signs for the translation from surface 2 back to surface 1′. The total value is then $4r/n'$ or $2r$. Multiplying a translation matrix with $2r$ in the lower left-hand corner by $-I$ gives (102), as expected. A similar result comes out of Figure 1.21, since $f' = r/2$ and the total translation is then r. On the other hand, the next two examples involve a lens and a mirror that are far apart. The translations from lens to mirror and back again cancel and we are left with a system matrix $-I$ that reverses the direction of the ray.

1.14 THE DESIGN OF A ZOOM LENS

To design an optical system, start with the simplest components: the thin lens, the reflector, or some combination of the two. If the design cannot be made to meet the requirements specified, there is no point in refining it by using the detailed specifications (radii, thicknesses, and indices) of the lenses. As an example of this kind of design using thin lenses, consider a camera lens that has been focused on an object. We wish to change the size of the image without affecting its sharpness. The optical system that meets these conditions is called a *zoom lens*. We realize that the minimum number of lenses needed is three; a two-lens system would have a variable magnification when the separation of the lenses is altered, but the image would not remain in focus because the Gaussian constants will also change. A three-lens system can be designed by trial-and-error methods, but it often pays to search the literature for ideas on how to start. Such a search resulted in the discovery that the matrix method has been extended by K. Halbach, *American Journal of Physics* 32, 90 (1964), and then used to design a zoom system. His procedure (which is somewhat complicated) is to define a *focal plane matrix* $S_{F'F}$ for a thin lens of focal length f' by using the translation matrices that connect V_1 with F and V_2 with F′. The three matrices to be multiplied are

$$S_{F'F} = \begin{pmatrix} 1 & 0 \\ l'_F & 1 \end{pmatrix} \begin{pmatrix} b & -a \\ -d & c \end{pmatrix} \begin{pmatrix} 1 & 0 \\ -l_F & 1 \end{pmatrix} \tag{103}$$

and, when (47) and (48) are used, we obtain

$$S_{F'F} = \begin{pmatrix} 0 & -a \\ 1/a & 0 \end{pmatrix} = \begin{pmatrix} 0 & 1/f' \\ -f' & 0 \end{pmatrix} \tag{104}$$

PROBLEM 42

Show that the focal plane matrix gives the correct locations of the cardinal points.

Extending this to two lenses with the object-space focal point F_2 of the second lens at a distance d_1 to the right of the image-space focal point F_1' of the first lens, their focal plane matrix is

$$S_{F'_2F_1} = \begin{pmatrix} 0 & 1/f_2' \\ -f_2' & 0 \end{pmatrix}\begin{pmatrix} 1 & 0 \\ d_1 & 1 \end{pmatrix}\begin{pmatrix} 0 & 1/f_1' \\ -f_1' & 0 \end{pmatrix} = \begin{pmatrix} -f_1'/f_2' & d_1/f_1'f_2' \\ 0 & -f_2'/f_1' \end{pmatrix} \tag{105}$$

PROBLEM 43

Use the results of Problem 23 to locate the cardinal points of the two-lens system.

A third lens can be added, with the distance between the focal points of lens 2 and lens 3 being designated as d_2. The combined focal plane matrix for all three lenses will be

$$S_{F_3'F_1} = \begin{pmatrix} -\dfrac{d_2f_1'}{f_2'f_3'} & \dfrac{d_1d_2}{f_1'f_2'f_3'} - \dfrac{f_2'}{f_1'f_3'} \\ \dfrac{f_1'f_3'}{f_2'} & -\dfrac{d_1f_3'}{f_1'f_2'} \end{pmatrix} \tag{106}$$

Halbach states that the condition of a variable image size without defocusing can be achieved if the focal points F and F′ of the system remain a fixed distance apart; he shows, by using the elements of the matrix in (106), that this requirement can be met if the following conditions hold.

$$f_1' = f_3', \qquad d_1 = -d_2 \tag{107}$$

That is, the first and third lenses are identical and the system is arranged as shown in Figure 1.23. The proof of (107) is rather complicated; it is more efficient to examine the design numerically. Let's try convex lenses with the two outer lenses having a focal length of 10 units and the inner lens with a focal length of 20 units. Let the outer lenses have a separation of 60 units, and place an object at 100 units from the first surface. We then have five matrices to multiply, and we shall use (64) and (25) rather than Halbach's form. There will be three thin-lens matrices separated by two translation matrices. The procedure for the numerical multiplication is the same as was used in connection with the Tessar. When lens 2 is moved from the center position to a distance of 5 units from this lens 1, the computation produces the results shown in Table 1.3. This table indicates that there is a reduced, real image about 10 units to the right of lens 3, that its magnification varies over a range of about two and one-half, and that the focal points F and F′ of the zoom system are a fixed distance apart. Note that this distance is equal to the separation of lenses 1 and 3, plus twice the focal length of the end lenses. This result can be shown to hold true in general. However, there is a flaw in the design because the image position is not fixed; it shows a variation of about 20%. A second calculation for the motion of lens 2 toward lens 3 gave similar results, and with approximately the same focusing error. To correct for this, it is necessary to allow one of the outer lenses to move. In most systems, it is convenient

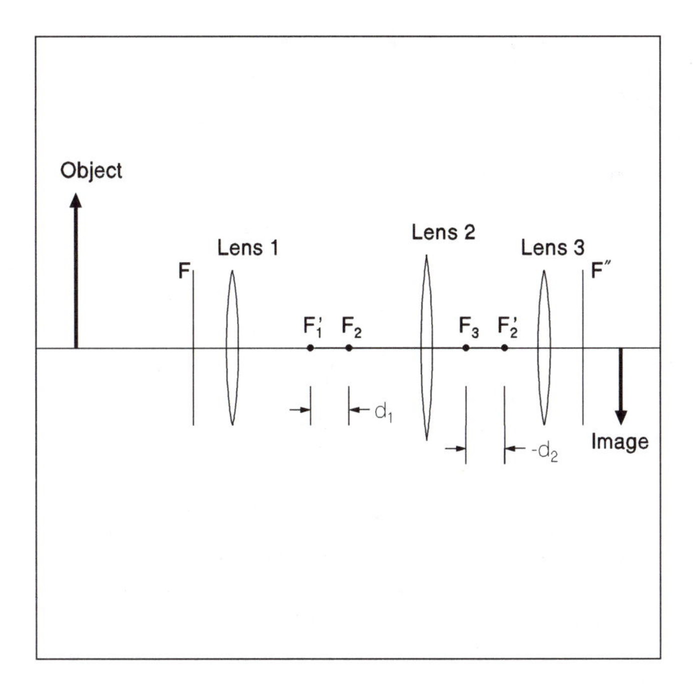

Fig. 1.23

Table 1.3 Program Output for Preliminary Design of Zoom System

Spacing of Lenses 1 and 2	Magnification	Image Position	I_F	I_F¢
30	0.056	10.278	–10.0	10.0
25	0.052	11.419	–8.8	11.2
20	0.043	12.174	–8.0	12.0
15	0.035	12.511	–7.6	12.4
10	0.027	12.568	–7.5	12.5
5	0.021	12.480	–7.6	12.4

to have the last surface remain at a fixed distance from the film or the detector; therefore we move lens 1 to refocus. Some additional steps in the computer program show that a small adjustment of lens 1 will ensure that the generalized Gaussian lens equation will be satisfied for each position of lens 2 as it moves towards lens 3. Since the total object-image distance for lens 2 at its center starting position is 100 + 60 + 10.278, the program was designed to move lens 3 until this distance was 170.3 units for all positions of lens 2. The necessary small motion of lens 1 as well as the simultaneous motion of lens 2 is shown in Table 1.4. Although this design, which is based on the properties of thin lenses, cannot be expected to give a high-quality image, it might

Table 1.4 Program Output for Final Design of Zoom System

Spacing of Lenses 1 and 2			
Initial	**Final**	**Magnification**	**Object-Image Distance**
35	35.7	0.05	170.31
40	40.8	0.04	170.35
45	45.6	0.04	170.35
50	50.3	0.03	170.25
55	55.2	0.02	170.40

still have applications involving the detection of light. To reduce the design to practice, the simultaneous motion of lenses 1 and 2 is achieved by mounting these components in a cylinder that can be rotated. A groove running from the center to the end where lens 3 is located has the form of a helix with uniform pitch, while a nonuniform groove takes care of the smaller amount of motion required for lens 1.

PROBLEM 44

Verify Tables 1.3 and 1.4 by writing the appropriate programs.

1.15 RETROFOCUS LENSES

A problem in designing a reflex camera is finding room for the mirror to swing out of the way when an exposure is made. The Tessar lens considered earlier has a focal length of 5.07 cm, which is typical of cameras using 35-millimeter film, but the distance from the last vertex to the focal point F′, called the *back focal length*, is only 4.40 cm, which is somewhat cramping. Wide-angle lenses, of focal lengths 2.8 to 4.0 cm, are even more awkward, and it is necessary to use a *retrofocus* design. Problem 45 explains this term and shows how to design such a lens.

PROBLEM 45

(a) The film in a 35 mm camera has a frame size of 24×36 mm. A photographer is taking a picture of an adult at a distance of 10 feet. How much of the person appears on the film? Is a wide-angle lens necessary?

(b) A double concave lens, $f' = -3$ cm, is followed at a distance of 3 cm by a double convex lens for which $f' = 2$ cm. Do a scale ray trace of a convenient object, showing that the lens system has its unit point H′ and focal point F′ well to the right of V_2, so that the back focal length is substantially greater than f'.

1.16 TELESCOPES AND MICROSCOPES

Telescopes are instruments that produce images of distant objects. It is generally believed that they magnify, but this is not usually the case, and the magnification m is not a very useful concept, as we shall demonstrate with an example. Suppose we wish to observe what goes on inside a vacuum system. Let the object be a disk, 1 inch or 25 mm in diameter, undergoing a coating process. If we look at the disk before it is placed in the chamber, we can see it most clearly if it is 10 in. or 254 mm—the *least distance of distinct vision*—from our eye. Then the half-angle subtended at the eye is arc tan (12.5/254) = 3°. At this distance, the eye can resolve about 6.5 cycles/mm, where a cycle consists of one black bar and one white bar of a periodic pattern. This is equivalent to 1/13 mm as the smallest visible detail, or about 75 μm. (Try this for yourself: make two marks 1 mm apart, and put 10 pinholes between them.) Now place the disk in a vacuum chamber that is so large that the disk is 6 m from the observer and has to be observed with a telescope. Let the telescope have two lenses: an *objective* with a focal length f'_1 = 120 cm, and an *eyepiece* for which f'_2 = 8 cm, with a spacing of 156 cm. Multiplying the thin lens matrices, we find that $a = -7/240$, $b = -37/2$, $c = -3/10$, and $d = -156$, giving a value of m exactly equal to –1, using $m = b + at$. The half-angle at the object without the telescope is arc tan (12.5/6000) = 0.01° and, with it, it is 3°, so that it *appears* much larger even though object and image are the same size. Since the magnification m does not really indicate the function of the telescope, we introduce a *magnifying power P*, defined as the ratio of half-angles with and without the telescope; P = 30.0 in this example, which is more meaningful than $m = -1$. Note that P bears no direct relation to m or to the angular magnification μ. Another example is the simple magnifier or *loupe*. Let us examine the disk in detail with a lens of focal length f' = 2.5 cm, placing the object so that the image is erect, virtual, and at a distance of –25.4 cm. Then the object distance, from (74), is $s = 2.5(-25.4)/(-25.4 - 2.5) = 22.8$ mm, the magnification is $m = -s'/s = 254/22.8 = 11$, and we obtain

$$P = \arctan(12.5 \times 11/254)/\arctan(12.5/254) = 10.1$$

so that m and P are almost numerically equal. Many telescopes consist of two thin lenses with a separation specified as

$$t'_2 = f'_1 + f'_2 \tag{108}$$

PROBLEM 46

(a) Show that the system matrix for such a telescope is

$$S_{41} = \begin{pmatrix} -f'_1/f'_2 & 0 \\ f'_1 + f'_2 & -f'_2/f'_1 \end{pmatrix} \tag{109}$$

(b) Use this result and a diagram to demonstrate that the focal length of the telescope is infinite. This is the *afocal* property referred to earlier.

(c) The focal length of the objective is much greater than the eyepiece. Show that the telescope produces a *reduced* image. Show, however, that the image is brought closer to the telescope by an amount m^2, and the net result is a magnification of the object.

The magnification m that we have been using really should be called the *transverse magnification*; it tells what happens normal to the axis. There is also a *longitudinal magnification* m' measured along the instrument axis.

PROBLEM 47

Use (44) and (109) to show that

$$m' = m^2 = 1/\mu^2 \tag{110}$$

The consequence of this is rather surprising. If we view a distant object with a telescope, we see a virtual image that is *smaller* than the object by a factor of m, but the longitudinal magnification brings it m^2 closer. It then appears larger because the angular magnification is increased by a factor of $1/m$. We can relate μ and P with a simple, approximate calculation. Suppose we have an object of height x at a distance z. For small angles, $\phi = x/(z + L)$ where L is the tube length. The angle for an image of height x' at 254 mm is $\phi' = x'/254$, so that $P = (x + L)/254\ \mu$. For large z, the longitudinal magnification $254/(z + L)$ is $254/z$, and $m' = 254/m = 1/\mu^2$. We thus obtain

$$P = \mu + \frac{L}{254\mu} \tag{111}$$

Binoculars are an example of an afocal optical system. A designation such as 7 × 50 means that $\mu = 1/m = 7$ and that the objectives have a diameter of 50 mm. A typical tube length is 180 mm. Hence $P = 7 + 180/(7 \times 254) = 7.1$, and there is virtually no difference between P and μ in this case. Two types of afocal telescopes in common use are the *Keplerian*, which has a converging eyepiece, and the *Galilean* or *Dutch*, which uses a diverging lens for the eyepiece. (Although Galileo gets the credit for this, he has written that the idea came from Holland.) This type is used in opera glasses, but it has an upper limit on attainable magnification; therefore, good binoculars are of the Keplerian type. These have a much greater tube length; a drawback that is eliminated by folding the optical path with reflecting prisms. Another kind of two-lens magnifying system is the *microscope*. The objective lens is very close to the specimen; the other lens is the eyepiece or *ocular*. The separation t'_2 is usually 160 mm and is known as the *tube length*. As a specific example, place a small object close to a lens of focal length 0.5 cm, and use an eyepiece of 2.0 cm focal length. It is possible for the image to be 100 cm below the eyepiece. This is done so that the eye is relaxed, as it will be when looking at a moderately distant image.

PROBLEM 48

Locate the object position and determine the magnification.

The magnification is a large number and quite meaningless; even the very best microscopes do not usually have values of m greater than about 1,200. As will be seen when we consider the wave nature of light, there is a natural limit on the attainable magnification of a microscope. It is the magnifying power P that is significant. To determine it, introduce the *optical tube length* L as the distance between F'_1 and F_2, or $L = 16 - f'_1 - f'_2$. For an image as produced by the objective to be close to F_2, the viewing angle (if small) is $\phi' = -x'/f'_2$, where x' is the image size. To a reasonable approximation, similar triangles generated by ray tracing will show that $x/f'_1 = x'/L$. Then $\phi' = -xL/f'_1 f'_2$. The angle with the unaided eye is $\phi = x/25.4$ giving

$$P = \frac{-25.4\ L}{f'_1 f'_2} \tag{112}$$

where all quantities are expressed in cm. For this example $P = (-25.4)(16 - 0.5 - 2.0)/(0.5 \times 2) = 343$. We can regard P as the product of two terms: $L/f'_1 = 13.5/0.5 = 27$ and $25.4/f'_2 = 25.4/2 = 12.5$ (approximately). The objective would then be marked as 27× and the eyepiece as 12.5×, with the combined magnification being simply the product. In practice, microscope components are more regular in their identification: 5×, 10×, 50×, and so on.

PROBLEM 49

Show that the focal length of the microscope is

$$f' = -f_1 f_2 / L \tag{113}$$

1.17 Summary of Design Procedure

Paraxial matrix optics has been developed and applied to some typical systems. Knowing the specifications of the lenses and mirrors, we can determine the location of the cardinal points and predict the behavior of the light rays. We can also use thin-lens matrices to start the design process. The first step will be a determination of how the object and image are related and the kinds of components necessary to obtain this relationship. Paraxial optics assumes that all components are perfect, so the next step will be to to eliminate the paraxial approximation and see exactly how the image is degraded by real components. This is the subject of the next few chapters. By way of review and summary, let us list the steps in the complete paraxial design or analysis of an optical system, no matter how simple or complex.

1. Start with the specification of radii or curvatures, refraction indices, and spacings. Compute the value of k at each refracting surface and the value of t'/n' for each translation.
2. Start the determination of the system matrix by multiplying the first translation matrix by the first refraction matrix and checking that $bc - ad = 1$ for this prod-

uct. Multiply this result by each succeeding matrix in the proper order and again check the determinant after each step.

3. Specify the object position and locate the image position using the Gaussian constants and the generalized Gaussian lens equation.
4. Calculate $1/m$ and then confirm this value by calculating m, using equations (41) and (42).
5. Locate the cardinal points and draw a ray tracing diagram to scale, using the focal and unit points. Then add the third set of rays, ensuring that they are parallel.
6. Compare the size, position, and orientation of the image on the diagram with the original calculations.

If each of the cross-checks specified in this procedure is found to be valid, then you can have confidence in the correctness of your design or analysis.

PROBLEM 50

A *Barlow* lens is a negative lens placed close to the objective lens of a telescope on the eyepiece side. Show that it increases the angular magnification by a factor $M = f'_1/(f'_B - d)$, where f'_1 is the focal length of the objective, f'_B is the focal length of the Barlow, and d is their separation.

PROBLEM 51

(a) An *autocollimator* is an instrument that can be used to find the location of the focal point F of a lens. A pinhole is placed in front of a bulb (Figure 1.24) and the lens and mirror are adjusted

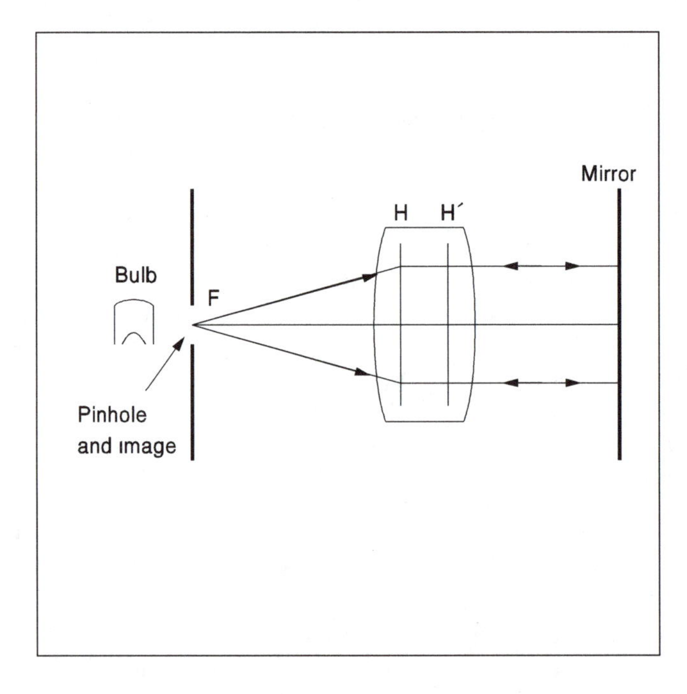

Fig. 1.24

so that the image formed is in focus at the pinhole. Explain why the distance from lens to pinhole is l_F.

(b) The same instrument can be used to locate the unit point H. Add a *nodal slide* (Figure 1.25), which is a lens holder that can move back and forth along the z-axis and can also be rotated about an axis perpendicular to the drawing. Show by a diagram that the image of the pinhole will not move when the slide is rotated if the axis is precisely at H. (*Hint*: Use elementary ray tracing and the binomial approximation.)

PROBLEM 52

The Gaussian constants can be measured experimentally by plotting the magnification as a function of object or image distance. Show that the curve of m vs. t' permits a determination of a and c, while the curve of $1/m$ vs. t determines a and b.

PROBLEM 53

A transparent pipe has an inner diameter of 6 in. and a wall thickness of 1 in. The index of refraction of a liquid flowing through the pipe is 4/3, and for the glass it is 3/2. It is well known to chemical engineers that the liquid will appear to be magnified.

(a) Prove that the magnification is precisely equal to the index of the liquid and is independent of the pipe dimensions and the index of its wall.

(b) Show that object and image coincide.

(c) Verify your calculations by tracing three sets of rays.

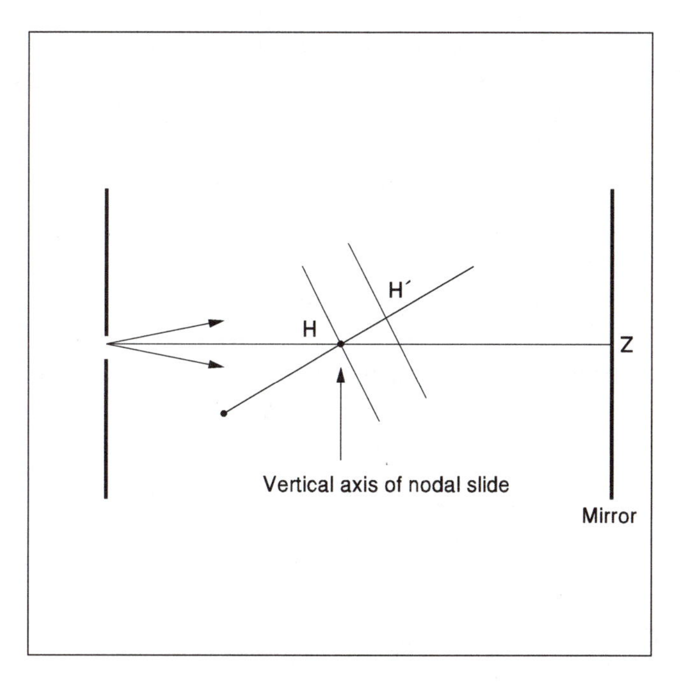

Fig. 1.25

1.18 Design Example

1.18.1 Objective

The goal of this project is the preliminary design of a small multipurpose pocket instrument.

a. Choose a lens that will provide a magnification of 5× ($m = 5$) for an object at a distance of 4 cm.

b. Let the object distance be raised to 200 cm. Design a small telescope by using the lens of part (a) and adding an eyepiece at a spacing of 4 cm to obtain a magnification of 3×.

c. Convert the telescope of part (b) into a jeweler's loupe by placing a third lens in contact with the objective. The working distance should be 2 cm and the magnification is 25×. An instrument that performs these three functions (and more) is the Emoskop, made by Siebert of Wetzlar, Germany.

1.18.2 Procedure

a. For a thin lens

$$a = 1/f' \quad b = c = 1 \quad d = 0$$

By (42)

$$1/5 = 1 + (-4)a \quad a = 1/5 \quad f' = 5 \text{ cm}$$

b. Equation (71) of Problem 23 gives expressions for a and b for two thin lenses at a separation of t_2'. Using these in the formula for $1/m$, and solving gives

$$f_2' = [t - (tt_2'/f_1') - t_2']/[(1/m) - 1 - (t/f_1')]$$

Using

$$t = -200 \quad t_2' = 4 \quad f_1' = 5 \quad m = 3$$

gives

$$f_2' = -1.1$$

A double concave lens with a focal length of −1 cm is available commercially. This instrument is a Galilean telescope.

c. Solving now for f_1'

$$f_1' = [t - tt_2'/f_2']/[(1/m) - 1 + (t_2' - t)/f_2']$$

from which $f_1' = 1.45$. For two thin lenses in contact

$$1/1.45 = 1/5 + 1/f_3'$$

so that the focal length of the third lens is $f_3' = 2$ cm.

CHAPTER 2

Nonparaxial Meridional Optics

2.1 Exact Meridional Ray Tracing

When we traced rays by using the cardinal points and planes, we knew that the image point P′ produced by an object point P was the destination of every ray leaving P. This was a consequence of the paraxial approximation. We now ask what happens when we are dealing with rays that have large inclinations with respect to the symmetry axis. Such rays will not meet at P′ and the image will no longer be a sharp point. This defect in the image is called a *lens aberration* or a *geometrical aberration*. We shall show that there are five distinct kinds of aberrations; let us start with the simplest.

Figure 2.1 shows the first surface of a lens. A ray leaves the object point P and strikes surface 1 at the point P_1. Rays that lie completely in the plane ZOX are said to be *meridional*; this is the plane that passes through the symmetry axis, just as meridians on the earth's surface are determined by planes through the north-south axis. We shall first consider the translation process, which will be expressed as a vector $\mathbf{T}_1$. We want to determine the matrix that describes this translation. Instead of specifying the inclination of a ray by the angle α that it makes with OZ, as was done for paraxial rays, we shall use the direction-cosines L, M, and N with respect to the three axes. For the time being, the direction-cosine M is not needed, but it will be used later for three-dimensional problems. The direction-cosines L and N are defined in terms of the components of $\mathbf{T}_1$ as

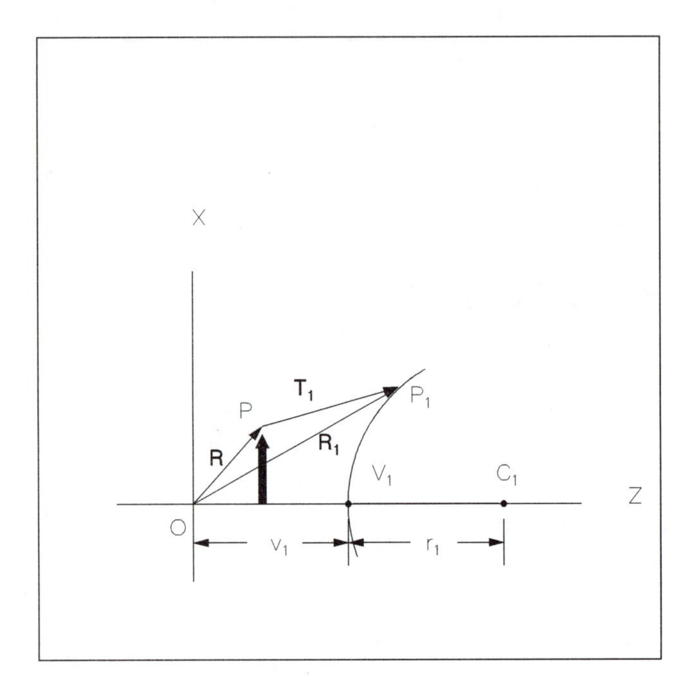

Fig. 2.1

$$L = T_{1x}/T_1, \qquad N = T_{1z}/T_1 \tag{114}$$

The location of the object point P, with coordinates x,y, and the intersection point P_1, at x_1, y_1, are also designated as vectors. The three vectors shown in the figure are related by

$$\boldsymbol{R}_1 = \boldsymbol{R} + \boldsymbol{T}_1 \tag{115}$$

Taking the dot or scalar product of each side with itself, we obtain

$$R_1^2 = R^2 + T_1^2 + 2\boldsymbol{R}\cdot\boldsymbol{T}_1 \tag{116}$$

Inserting the components of the two vectors into the final term on the right yields

$$R_1^2 = R^2 + T_1^2 + 2T_1(zN + xL) \tag{117}$$

We now introduce the properties of surface 1. The equation of a circle with its center at the origin is

$$z^2 + x^2 = r^2 \tag{118}$$

in the coordinate system we are using. If we move the center to C_1, with the vertex at a distance v_1 from the origin, the equation for surface 1 becomes

$$(z_1 - v_1 - r_1)^2 + x_1^2 = r_1^2 \tag{119}$$

This may be written as

$$z_1^2 + x_1^2 = 2z_1(r_1 + v_1) - v_1^2 - 2v_1r_1 = R_l^2 \tag{120}$$

Combining this with equation (117) and eliminating z_1 using

$$z_1 = z + T_1N \tag{121}$$

gives

$$T_1^2 + 2T_1[N(z-r_1-v_1) + Lx] + (z-v_1)^2 + x^2 + 2r_1(v_1-z) = 0 \tag{122}$$

This is a quadratic equation of the form

$$T_1^2 + BT_1 + C = 0 \tag{123}$$

and it has two roots. However, we shall show that the only meaningful solution is

$$T_1 = \frac{-B - \sqrt{B^2-4C}}{2} \tag{124}$$

but we must first recast this equation because it cannot handle flat surfaces (the value $r = \infty$ causes difficulties in a computer program). We do this by multiplying top and bottom by the quantity

$$-B + \sqrt{B^2 - 4C} \tag{125}$$

to obtain

$$T_1 = \frac{2C}{-B + \sqrt{B^2 - 4C}} \tag{126}$$

Then we introduce the curvature $c_1 = 1/r_1$ and let

$$E = Bc_1/2 = c_1[(z - v_1)N + xL] - N \tag{127}$$

and

$$F = Cc_1 = c_1[(z-v_1)^2 + x^2] - 2(z-v_1) \tag{128}$$

obtaining

$$T_1 = \frac{F}{-E + \sqrt{E^2 - c_1F}} \tag{129}$$

Note that, when the curvature $c_1 = 0$, as for a plane interface, then

$$T_1 = \frac{2(v_1-z)}{N + N} \tag{130}$$

or

$$N = (v_1 - z)/T_1 \tag{131}$$

which we recognize as the definition of the direction-cosine N of a ray that strikes a plane surface passing through the first vertex. This confirms the correctness of our choice of sign in the solution of the quadratic equation 123. The coordinates of P_1 may now be found from (121) and the corresponding equation for x_1. However, it is customary to shift the origin after each translation for this reason: the distances along the axes of most optical systems are usually large compared to those in the x and y directions. In order to avoid having to deal with large and small numbers in the same computer calculation, after each translation we move the origin to the vertex at which the next refraction is going to occur. In the present case, this is accomplished by changing (121) to

$$z_1 = z + T_1 N - v_1 \tag{132}$$

and the other equation is

$$x_1 = x + T_1 L \tag{133}$$

Since the angles or the direction-cosines remain unchanged during a translation, these two equations can be put into matrix form as

$$\begin{pmatrix} n_1 N \\ z_1 + v_1 \end{pmatrix} = \begin{pmatrix} 1 & 0 \\ T_1/n_1 & 1 \end{pmatrix} \begin{pmatrix} n_1 N \\ z \end{pmatrix} \tag{134}$$

and

$$\begin{pmatrix} n_1 L \\ x_1 \end{pmatrix} = \begin{pmatrix} 1 & 0 \\ T_1/n_1 & 1 \end{pmatrix} \begin{pmatrix} n_1 L \\ x \end{pmatrix} \tag{135}$$

The 2×2 matrix in these equations is the *nonparaxial* or *exact translation matrix*. This indicates that the nonparaxial translation matrix is a more general form of the paraxial matrix we have been using; the constant translation t_1 as measured along the z-axis is replaced by a variable distance T_1 measured along the ray. It is an exact quantity and changes from ray to ray.

The form of the exact refraction matrices can be obtained from Figure 2.2, which shows the incoming ray at surface 1. This ray is taken to be a vector $\mathbf{n}_1$ with magnitude equal to the index of refraction on the left. Then the refracted ray is designated as $\mathbf{n}_1'$ in a similar way.

Define a quantity K_1, called the *refracting power* or *skew power*, in terms of these two rays by the relation

$$K_l \mathbf{s}_1 = c_1(\mathbf{n}_1 - \mathbf{n}'_1) \tag{136}$$

where c_1 is the curvature and $\mathbf{s}_1$ is a unit normal vector at the surface. This definition

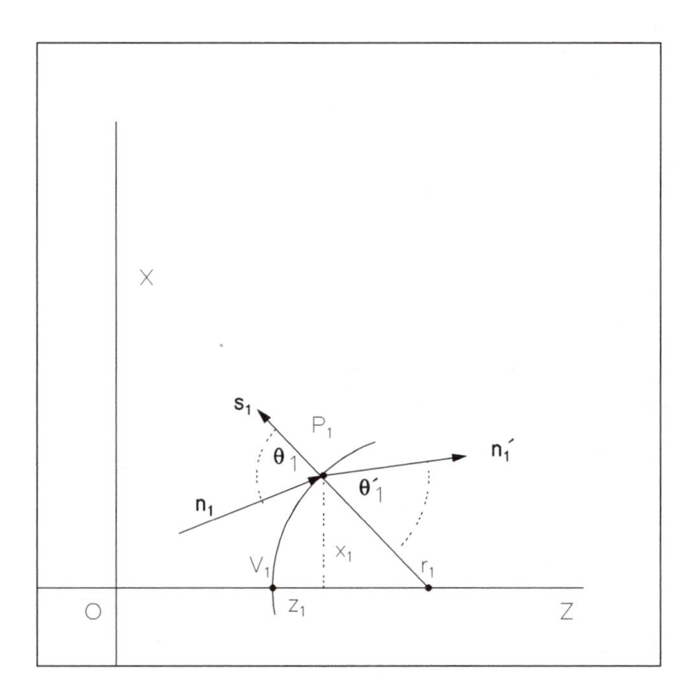

Fig. 2.2

implies that the unit normal and the vector $(\mathbf{n_1} - \mathbf{n_1}')$ must be parallel. Problem 54 shows the consequence of this condition.

PROBLEM 54

A ray strikes surface 1 of a lens at an angle of 45°. Show that the angle of refraction is slightly less than 30° for glass with an index of 1.500. Then use this result to draw a scale diagram showing the incident and refracted rays, put in the negative of the refracted ray, and demonstrate that the combined vector on the right of (136) is parallel to the normal at the surface. Show on your diagram that this condition is a consequence of Snell's law.

The scalar product of both sides of equation (136) with the unit normal vector gives

$$K_1 = c_1[n_1 \cos(180-\theta_1) - n_1' \cos(180-\theta_1')] \tag{137}$$

or

$$K_1 = c_1(n_1' \cos \theta_1' - n_1 \cos \theta_1) \tag{138}$$

For small angles, the paraxial approximation causes this expression to reduce to k_1, as defined by equation (17), so that we again have the situation where the originally constant quantity now becomes dependent on the angle at which the ray strikes the lens. To determine K_1, we need the cosines of the angles in Figure 2.2. The scalar product

$$\boldsymbol{n}_1 \cdot \boldsymbol{s}_1 = n_{1z}s_{1z} + n_{1x}s_{1x} = n_1 N_1 \frac{z_1 - r_1}{r_1} + n_1 L_1 \frac{x_1}{r_1} \tag{139}$$

can also be expressed as

$$\boldsymbol{n}_1 \cdot \boldsymbol{s}_1 = n_1 \cos(180° - \theta_1) = -n_1 \cos\theta_1 \tag{140}$$

where Figure 2.2 shows that the two vectors involved in this scalar product meet head to tail. The components of each of these two vectors are obtained by multiplying their magnitudes by the direction-cosines with respect to the axes. Combining these equations and using the direction-cosines for the unit normal as indicated by the figure, we obtain

$$\cos\theta_1 = N_1 \frac{r_1 - z_1}{r_1} - L_1 \frac{x_1}{r_1} \tag{141}$$

or

$$\cos\theta_1 = N_1(1 - c_1 z_1) - L_1 x_1 c_1 \tag{142}$$

where z_1 is measured from V_1, as we just mentioned. Having found the incident angle, we can find the angle after refraction with Snell's law, and equation (138) becomes

$$K_1 = c_1[n_1'\sqrt{1-(n_1/n_1')^2(1 - \cos^2\theta_1)} - n_1\cos\theta_1] \tag{143}$$

Note that n_1' is kept outside the radical; for mirrors, the signs on the indices are crucial, and they would be lost if we didn't write equation (143) as shown. Signs for the angles are also crucial—we will consider them shortly. We now have all the information needed to calculate K_1; to use this quantity, write (136) as two scalar equations. These are

$$K_1 x_1 = L_1 n_1 - L_1' n'_1 \tag{144}$$

and

$$K_1(c_1 z_1 - 1) = c_1(N_1 n_1 - N_1' n_1') \tag{145}$$

Or in matrix form

$$\begin{pmatrix} n_1'N_1' - K_1/c_1 \\ z_1' \end{pmatrix} = \begin{pmatrix} 1 & -K_1 \\ 0 & 1 \end{pmatrix} \begin{pmatrix} n_1 N_1 \\ z_1 \end{pmatrix} \tag{146}$$

and

$$\begin{pmatrix} n_1'L_1' \\ x_1' \end{pmatrix} = \begin{pmatrix} 1 & -K_1 \\ 0 & 1 \end{pmatrix} \begin{pmatrix} n_1 L_1 \\ x_1 \end{pmatrix} \tag{147}$$

These equations show that the nonparaxial refraction matrix is a generalization of the paraxial matrix, to which it reduces for small angles, in the same way as the translation matrix.

2.2 NUMERICAL RAY TRACING

Consider now a double convex lens with radii of 50 units, a thickness of 15 units, and an index of 1.5. Let a meridional ray be parallel to the z-axis and 2 units above it. The BASIC program for performing the calculation follows. The first part specifies the object location and the properties of the lens.

```
DIM T(12), C(12), N(12), NP(12)
T(1) = 200: T(2) = 15: T(3) = 47.368
C(1) = .02: C(2) = -.02: C(3) = 0
N(1) = 1: N(2) = 1.5: N(3) = 1
NP(1) = 1.5: NP(2) = 1: NP(3) = 1
```

The next step is to locate the paraxial focal point F', using equations (29) and (31) to find the ratio c/a. The ray will be traced to this plane. The final portion of the program starts by indicating the location and slope of the ray. All the constants are incorporated into a procedure that is a direct application of the equations just derived. This part of the program is

```
X = 2: Z = 0
EL = 0: EN = SQR(1 - EL*EL)
FOR J = 1 TO 3
ZV = Z - T(J)
E = C(J) * (X * EL + ZV * EN) - EN
F = C(J) * (X * X + ZV * ZV) - 2 * ZV
SD = F / (SQR(E * E - F * C(J)) - E)
X = X + SD * EL
Z = Z + SD * EN - T(J)
OP = 1 - C(J) * Z
COSTHETA = EN * OP - C(J) * EL * X
BC = N(J) * N(J) * (1 - COSTHETA * COSTHETA) / NP(J) / NP(J)
KOVERC = NP(J) * SQR(1 - BC) - N(J) * COSTHETA
K = C(J) * KOVERC
EL = (N(J) * EL - K * X) / NP(J)
EN = (N(J) * EN + KOVERC * OP) / NP(J)
PRINT Z, X, EN, EL
NEXT J
```

The dimension statement permits up to 12 translations T, curvatures C, incident indices N, and refracted indices NP. The first translation T(1) starts the ray 200 units to the left of vertex 1. Observe that T(1) is taken to be the positive distance from the object at the origin to the lens. This is opposite to the convention we used in chapter 1;

we do it because it simplifies the programming. The next translation T(2) is the lens thickness, and the translation T(3)—the distance from vertex 2 to the image plane—is the quantity l_F', determined as in Problem 22. The two radii are expressed in terms of their reciprocals, although they could just as well be written as C(1) = 1/50, C(2) = –1/50. The indices n and n' are listed separately but, since $n_1' = n_2$, it would be feasible in a long program to set up a procedure that would get any n' from the associated n. The ray tracing procedure starts with the coordinates $z = 0$ and $x = 2$ and the direction-cosines $N = 1$ and $L = 0$ for a ray parallel to the z-axis. Then there is a FOR, NEXT loop with three steps: step 1 takes the ray from object to vertex 1; step 2 takes it to vertex 2; step 3 traces the ray out to the image plane. The translation from object to vertex 1 is specified by the slant distance T_1, symbolized by SD. We calculate it from (129), using E and F as given by (127) and (128). Then the coordinates of the ray at surface 1 are determined from the nonparaxial matrix equations (134) and (135). Note that the new value of z is expressed with respect to the first vertex. These quantities are next used in the refraction computation. We find the cosine of the incident angle from (141). This value is substituted into (143) to obtain K_1/c_1 (designated KOVERC), and the next two steps find the direction-cosines of the emerging ray as specified by the matrix refraction equations. This process is repeated at the second vertex and at the paraxial focal plane. Note that, at F′, we compute refraction angles even though the index is the same on both sides of the plane. This is unnecessary and could be eliminated, but the program is simpler when it is included. The values of z,x and N,L—given just below—are printed after each of the three steps. We find that the ray crosses the paraxial focal plane at 0.004 units below F′. This is a small, but noticeable, defect in the lens called *spherical aberration*. By changing the initial position of the incoming parallel ray, we can see from Table 2.1 that a ray at a distance of 20 units from the axis will show a noticeable increase in this defect, while a ray that is very

Table 2.1 Ray Tracing Results (Paraxial focal point: 47.368)

	Z	X	N	L
$x = 2$				
	0.040	2.000	0.999	–0.013
	–0.032	1.800	0.999	–0.038
	0.000	–0.004	0.999	–0.038
$x = 20$				
	4.174	20.000	0.990	–0.141
	–3.747	18.991	0.890	–0.456
	0.000	–7.197	0.890	–0.456
$x = 0.2$				
	0.000	0.200	1.000	–0.001
	0.000	0.180	1.000	–0.004
	0.000	0.000	1.000	–0.004

close to the axis (x = 0.2) will be effectively paraxial. One important result is the sign of *N*. The initial value of the *z*-axis direction-cosine is 1.0, and it decreases slightly as the ray is bent down toward the axis, but it stays positive as long as the ray is moving to the right. On the other hand, *L* goes from zero to a small negative number, as it should. However, if the ray is refracted upward, then *L* will also be positive, so that you can immediately tell from its sign which way the ray was bent.

PROBLEM 55

Use the behavior of the slant distance T_1 as a criterion for deciding how close the rays have to be to the axis for the paraxial approximation to be valid.

2.3 A STUDY OF SPHERICAL ABERRATION

Although we have introduced spherical aberration through the use of a numerical computation, it is helpful to have a graphical representation. Figure 2.3 shows how rays that are parallel to the axis of a double convex lens are refracted as they travel toward the paraxial focal plane. The paraxial rays—those that are very close to the axis—will meet as expected at F′ (although this is difficult to see in this figure). The rays a little farther out—the *intermediate* rays—will cross the axis just a little to the left of F′, and those near the edge of the lens—the *marginal* rays—will fall very short, as indicated. Rotating this diagram about the axis, we realize that spherical aberration produces a very blurry circular image at the paraxial plane. The three-dimensional envelope of the group of rays produced by this rotation is known as the *caustic surface*, and the narrowest cross-section of this surface—a little to the left of the paraxial focal plane—is called the *circle of least confusion*. If this location is chosen as the image plane, then the amount of spherical aberration can be reduced somewhat. Other methods of improving the focusing will be considered later. Figure 2.3 was obtained by adding a graphics step to the program given just above. The BASIC language has a command

```
LINE (Z1, X1)-(Z2,X2)
```

that draws a line from the point with coordinates z_1, x_1 to the point with coordinates z_2, x_2 on the screen. We add this step at the end of the translation calculation, and—with a few other modifications—the program now becomes

```
XST = -25: DELX = 2:X0 = XST
10 X = X0: Z = 0
Z1 = Z:X1 = X: T1 = 0
EL = 0: EN = SQR(1 - EL*EL)
FOR J = 1 TO 3
ZV = Z - T(J)
E = C(J) * (X * EL + ZV * EN) - EN
F = C(J) * (X * X + ZV * ZV) - 2 * ZV
SD = F / (SQR(E * E - F * C(J)) - E)
X = X + SD * EL
```

```
Z = Z + SD * EN - T(J)
T1 = T1 + T(J)
Z2 = Z + T1: X2 = X
LINE (Z1, X1)-(Z2, X2)
Z1 = Z2: X1 = X2
OP = 1 - C(J) * Z
COSTHETA = EN * OP - C(J) * EL * X
BC = N(J) * N(J) * (1 - COSTHETA * COSTHETA) / NP(J) / NP(J)
KOVERC = NP(J) * SQR(1 - BC) - N(J) * COSTHETA
K = C(J) * KOVERC
EL = (N(J) * EL - K * X) / NP(J)
EN = (N(J) * EN + KOVERC * OP) / NP(J)
NEXT J
X0 = X0 + DELX
IF X0 <= -XST THEN 10
```

The figure also shows the lens, and this is created by considering a surface of radius r and with its center located at a distance r to the right of the origin. Its equation will be

$$(z - r)^2 + x^2 = r^2$$

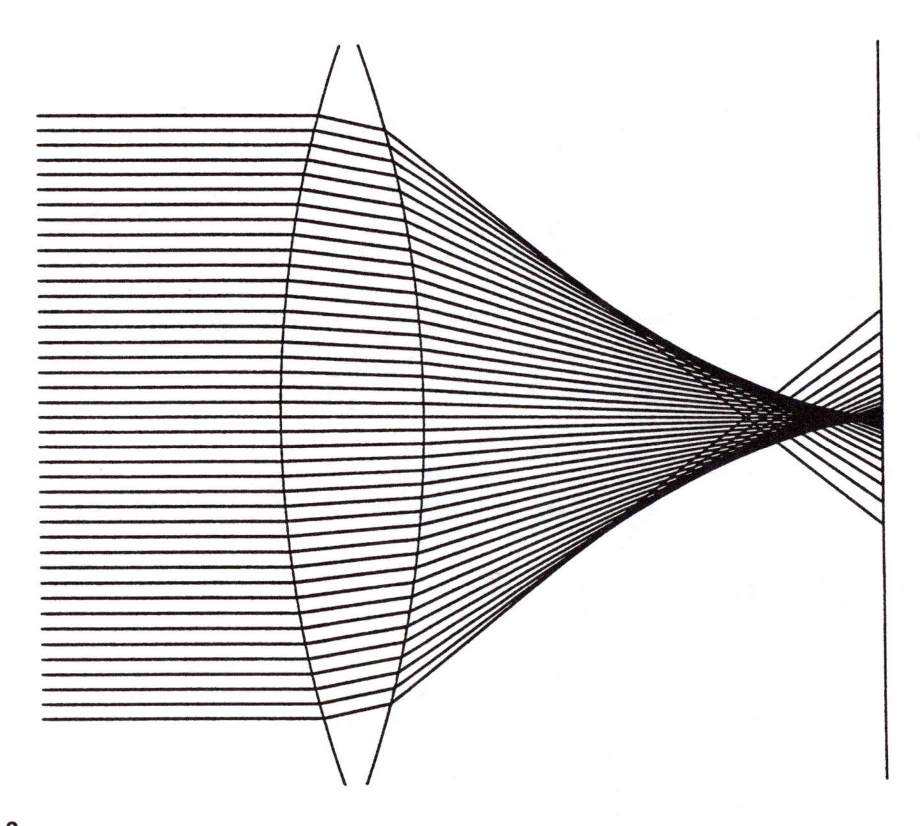

Fig. 2.3

or replacing r with $1/c$

$$z^2 - 2z/c + x^2 = 0$$

Solving for z, we get

$$z = \frac{cx^2}{1 \mp \sqrt{1 - c^2x^2}} \tag{148}$$

If the negative sign is used, we obtain $z = \infty$ for $c = 0$, so that the positive sign is the correct choice, and the equation for a circle that we shall use is

$$z = \frac{cx^2}{1 + \sqrt{1 - c^2x^2}} \tag{149}$$

Note that $z = 0$ for $c = 0$, which is correct for a vertical line through the origin. Thus, (149) automatically handles the image plane.

The program that draws the lens is

```
EDGE = -20
T1 = 0
FOR Q = 1 TO 3
T1 = T1 + T(Q)
X = EDGE
ARG = 1 - C(Q) * C(Q) * X * X
IF ARG < 0 THEN 22
Z = X * X * C(Q) / (1 + SQR(ARG)) + T1
22 X8 = X: Z8 = Z
20 X9 = X8 + 1
ARG = 1 - C(Q) * C(Q) * X9 * X9
IF ARG < 0 THEN 24
Z9 = X9 * X9 * C(Q)/(1 + SQR(ARG)) + T1
LINE (Z8, X8) - (Z9, X9)
24 X8 = X9: Z8 = Z9
IF X8 <= -EDGE THEN 20
NEXT Q
```

It should also be mentioned that a coordinate system for the monitor has to be specified at the beginning of the graphics procedure. This is accomplished by establishing a window with the command

```
WINDOW (60, -20)-(200, 20)
```

which sets the left-hand limit at 60 units, the right-hand limit at 200 units, and the lower and upper limits at –20 and 20 units, respectively.

PROBLEM 56

To draw the lens accurately, it is necessary to know the coordinates of the lens edges (the two intersections of the front and back surfaces).

(a) Show that the lens edge occurs at a distance from the axis given by

$$x_{max} = [2r_2 s_2 - s_2^2]^{1/2} = [2r_1 s_1 - s_1^2]^{1/2}$$

where s_1 and s_2 are the horizontal distances of the edge from the first and second vertex, respectively.

(b) Express s_1 and s_2 in terms of curvatures, and verify that your formulas give the correct results for a symmetrical lens.

(c) Suggest a way of using these results for a diverging lens, and devise a way of showing the flat edges.

Spherical aberration as presented in most optics texts is defined as the variation of focus with aperture, where *aperture* means the diameter of the section of the lens used to form the image. (See, for example, W. Smith, *Modern Optical Engineering*, 2d ed. McGraw-Hill, [1990]). This definition is inconsistent with the approach used here, and the source of the contradiction is the nature of paraxial optics. All the basic quantities of Chapter 1—focal length, magnification, cardinal points and planes, among others—are strictly paraxial concepts. When the paraxial approximation is no longer valid, we can no longer state that more than two rays from an object point will intersect, nor can we specify a unique image. With regard to magnification, for example, we would not know which image has to be used to measure this quantity. Therefore, we shall use a more appropriate definition of spherical aberration: *it is the failure of meridional rays to obey the paraxial approximation*. Figure 2.4 shows how to define spherical aberration quantitatively. Two kinds of aberration are specified in this figure. The place where the ray crosses the axis, lying to the left of F′, is at a distance,

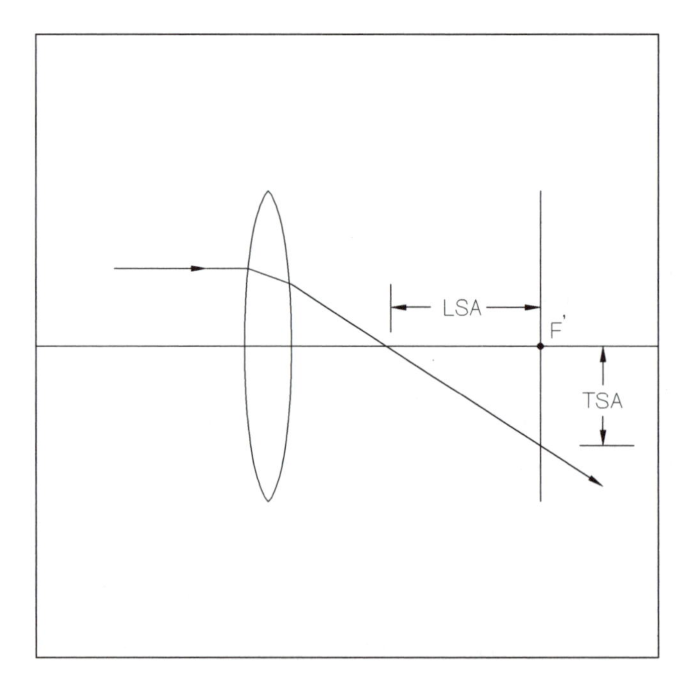

Fig. 2.4

measured from the paraxial focus, and called the *longitudinal spherical aberration* (LSA), while the distance below the axis and at the focal plane is the *transverse spherical aberration* (TSA). To reduce this aberration, note in Figure 2.3 that the amount of refraction at the two surfaces of the lens is unequal. Equalizing the refraction at the two surfaces will improve the situation. Suppose we alter the lens while keeping the index, thickness, and focal length constant; only the radii will change. This can be done if the new radii satisfy equation (59). The process of varying the shape of a lens while holding the other specifications constant is called *bending the lens*. This can be accomplished by a computer program, obtaining what are known as *optimal shape lenses*, which are commercially available. To study the effect of optimal shape lenses on spherical aberration, define the *shape factor* σ of a lens as

$$\sigma = \frac{r_2 + r_1}{r_2 - r_1} \tag{150}$$

from which $\sigma = 0$ for a symmetric lens ($r_1 = -r_2$). It is convenient to change a radius from the keyboard; the BASIC language has this capability. Figure 2.5a shows the screen for a symmetric lens, and Figure 2.5b illustrates the result of lens bending. These figures were obtained by using the Print Screen key. This lens has had the curvature of the first surface raised and the second surface has been flattened to keep the focal length constant. The legends printed on the screen are obtained by using the LOCATE command. For example, the legend FIRST RADIUS and the value of r_1 are produced by the BASIC steps

```
LOCATE 5, 50
PRINT "FIRST RADIUS = "; R(1)
```

where the coordinates 5, 50 mean 5 rows down from the top of the screen and 50 columns from the left. The interactive keyboard capability, which comes at the end of the program, appears as follows

```
DO
    key$ = INKEY$
LOOP WHILE KEY$ = ""
ky = ASC(RIGHT$(key$, 1))
SELECT CASE
    CASE 72: R(1) = R(1) + 5
    CASE 80: R(1) = R(1) - 5
    CASE 75: R(1) = R(1) + .5
    CASE 77: R(1) = R(1) - .5
END SELECT
```

The DO, LOOP WHILE loop acts to keep the program from going on to the next step, thereby keeping any messages on the screen. The INKEY$ function is used to detect which key is pressed to end this loop and continue with the program. It works by returning a two-byte string whose first byte is zero and whose second byte has a decimal value. These values, for the keys that are used, are Up Arrow (72), Down Arrow (80), Left Arrow (75), and Right Arrow (77). Other pairs that are conveniently arranged on the IBM keypad are Page Up (73), Page Down (81), Insert (82), Delete

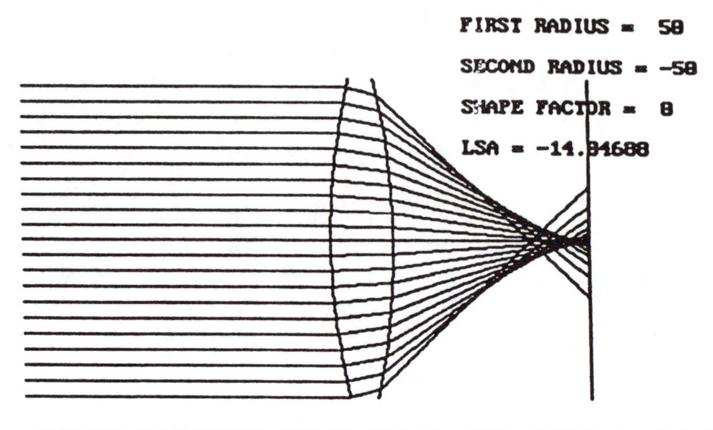

Fig. 2.5a

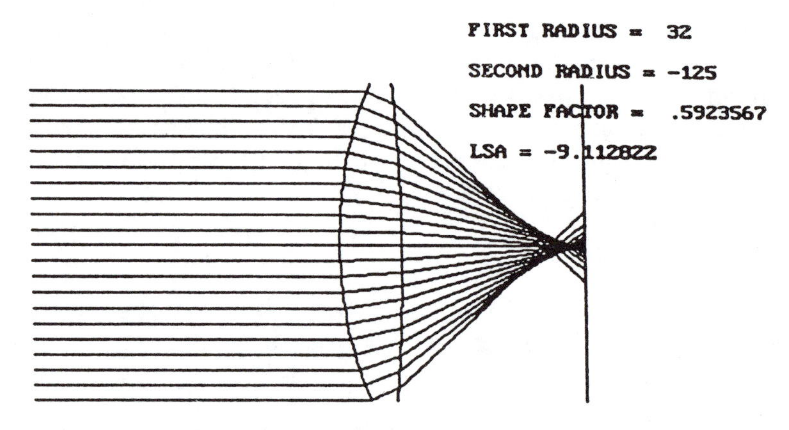

Fig. 2.5b

(83), Home (71), and End (79). In addition, the function keys F1–F10 (59–68) can be used. When the program leaves the loop by pressing an arrow key to make a change or pressing any other key to end the program, INKEY$ specifies a string variable key$ which is used with the RIGHT$ function. The notation RIGHT$(key$,1) means take the single rightmost character of key$, which in this case will be the second of the two bytes. Then the ASC function converts this into ASCII form, thus setting the variable key equal to 72, 80, 75, or 77, or to a completely different number. The next loop then takes the number corresponding to one of the arrows and changes the radius r_1 by the amount chosen. Then the screen should be cleared with CLS and the program restarted by a GO TO command which takes it back to just below the step where r_1 was originally given. If any other key but an arrow is pressed, the CASE ELSE command terminates the program.

PROBLEM 57

Write the program that will bend the lens of Figure 2.3 interactively from the keyboard, and plot LSA as a function of σ over a range wide enough to show the minimum.

2.4 A CLOSER LOOK AT RAY TRACING

The programs developed so far, when applied to new examples, may lead to error messages, which we shall show arise from the two square roots that are a part of these programs. The first one occurs in the calculation of the slant distance, and the second one is part of the refraction process. Whenever the argument of either of these radicals goes negative, the program stops.

PROBLEM 58

(a) Move the object in Figure 2.1 to the origin by letting $z = 0$, and show that the solution to equation (122) is

$$T_1 = N(v_1 + r_1) - Lx \pm \sqrt{[N(v_1 + r_1) - Lx]^2 - [x^2 + (v_1 + r_1)^2 - r_1^2]} \tag{151}$$

(b) Place an object of height x at the origin. Draw a ray leaving point P with positive slope α so that it strikes the front surface of a sphere at P_1 and passes through to strike the back surface at P_2. Then the slant distance T_1 has one of two possible values, depending on whether the lens is convex or concave. Drop a perpendicular from the center C_1 to the ray; draw the radii from C_1 to P_1 and P_2 (thus generating an isosceles triangle with the perpendicular down the middle); add a few more judiciously chosen construction lines to show geometrically that

$$T_1 = (v_1 + r_1)\cos\alpha - x\sin\alpha \pm \sqrt{r_1^2 - [x\cos\alpha + (v_1 + r_1)\sin\alpha]^2} \tag{152}$$

(c) Prove that (151) and (152) are the same equation.

It will be found when deriving (152) that the second term inside the radical on the right is the length of the perpendicular from the center of curvature to the ray. This line will be smaller than the radius if the ray crosses the circle twice. As the angle increases, the two intersection points will get closer, and they will merge when the ray is tangential to the circumference. Then the perpendicular is exactly equal to the radius and the radical vanishes. If the angle gets any larger, the ray completely misses the lens, the perpendicular exceeds the radius, and the radical becomes imaginary. We should therefore put an error-detecting mechanism into the program to avoid this possibility.

The other difficulty with imaginary quantities comes about from the physics of refraction. When a ray goes from air to water it is bent toward the normal; it is bent away from the normal when it goes from water to air. In this latter case, Snell's law tells us that, if the angle in water is large enough, the ray can emerge at right angles to the normal. This occurs when

$$n' \sin \theta' = 1 \sin 90° = 1 = n \sin \theta$$

or

$$\theta = \arcsin (1/n)$$

The value of θ corresponding to this phenomenon is known as the *critical angle*; for an air-water interface it is

$$\theta = \arcsin (1/1.33) = 48.8°$$

This is a rather large angle—for air-glass interfaces, it is about 45°; angles of this magnitude rarely occur in optical systems. However, the possibility has to be considered. When the critical angle is exceeded, the ray cannot emerge from the water; instead, it undergoes *total internal reflection*. That is, the air-water or air-glass interface acts like a mirror. The effect of this behavior on (143) can be seen by writing the quantity in the radical as $[1 - (n_1/n'_1)^2 \sin^2 \theta_1]$. It will become negative if

$$(n_1/n'_1)^2 \sin^2 \theta_1 > 1$$

Letting n_1 correspond to glass and n'_1 to air, this inequality states that the square root becomes imaginary when the critical angle is exceeded. This is another precaution that should be incorporated into ray tracing programs.

2.5 Spherical Aberration and Lens Design

Let us now consider what lens designers can do about spherical aberration. W. Brouwer, in his book *Matrix Methods in Optical Instrument Design* (W. A. Benjamin, 1964), gives the specifications shown in Table 2.2 for a cemented doublet. The front element of this lens by itself has a focal length of about 50, and the longitudinal spherical aberration varies as shown in Figure 2.6. The curve in Figure 2.6 can be approximately fitted by the equation

$$\text{LSA} = Ax^2 + Bx^4$$

where A and B are constants whose values are not needed. Let us require that the aberration vanish at the lens edge x_{max}, so that

$$A = -B\, x_{max}^{\;2}$$

Table 2.2 Specifications for a Cemented Doublet

r	*t′*	*n′*
61.070		
	4.044	1.56178
−47.1107		
	2.022	1.70100
−127.098		

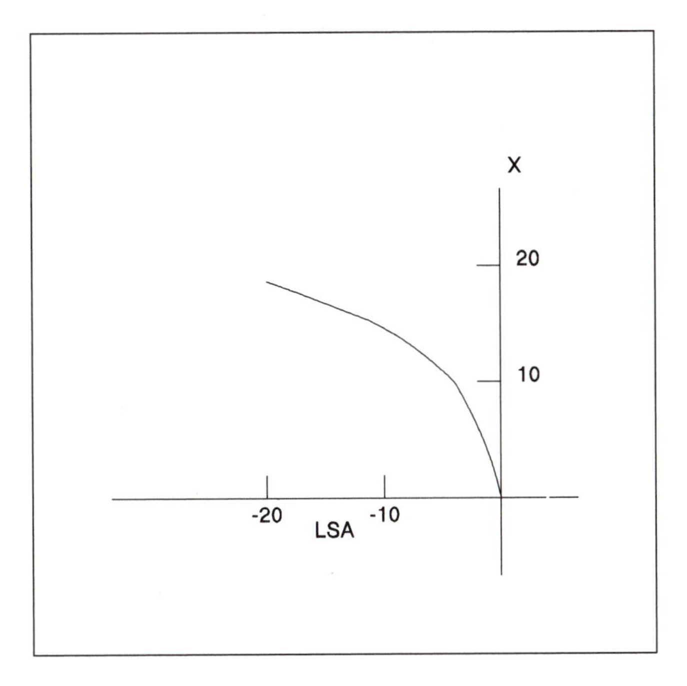

Fig. 2.6

and

$$\text{LSA} = B(-x_{\max}^2 x^2 + x^4)$$

Since the spherical aberration is zero on the axis and at the margin, it must have a maximum at some intermediate location. Differentiating and equating the derivative to zero shows that this value of x is 0.707 $x_{\max}$, or about three-quarters of the way between center and edge. The way that the designer arranged for the marginal rays to meet at the paraxial focus, rather than falling short, was to recognize that a diverging lens following the front element would refract the rays away from, rather than closer to, the axis. When this second component is added, we would expect the behavior shown in Figure 2.7, as taken from Brouwer. Note that the value of LSA for the doublet has been enormously reduced, as indicated by the horizontal scale change.

PROBLEM 59

Although Figure 2.7 correctly indicates that the LSA for the doublet vanishes on the axis and at the margin, and that it has a maximum at about 0.7 $x_{\max}$, it was found that the curve given could not be confirmed. Show that an increase in the magnitude of r_3 will correct this difficulty.

PROBLEM 60

It has been suggested by G. S. Fulcher, *Journal of Optical Society of America* 37, 47(1947), that a lens system consisting of a series of *meniscus* lenses (Figure 2.8) would have very low spherical

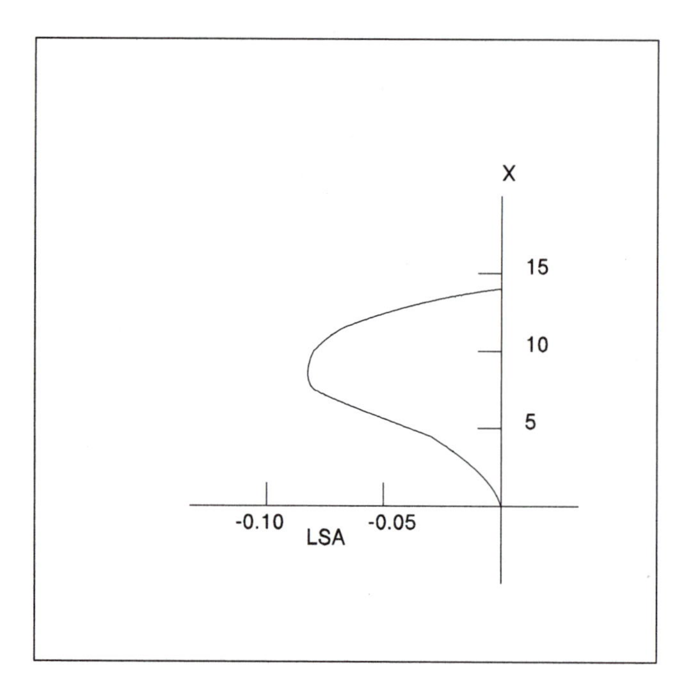

Fig. 2.7

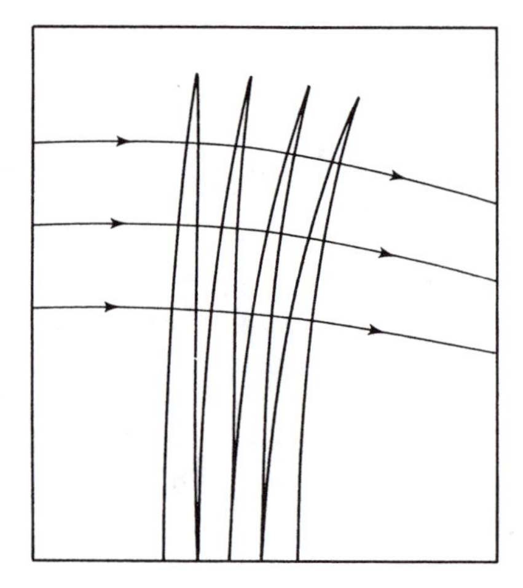

Fig. 2.8

aberration. These are spectacle lenses made of crown glass with an index of 1.5230. The idea he advances is that the paraxial rays should be subject to less refraction than the intermediate or marginal rays. By having the lenses in contact in the paraxial region, only surfaces 1 and 8 are active, but, at the edge, the refraction is spread over many interfaces. The eight radii of this lens system are, in the usual order, 20.92, ∞, 13.075, 34.8867, 8.7167, 14.9429, 6.9733, and 10.46. The thickness of all four lenses is 0.20, and they are in contact at the three common vertices.

(a) Find the LSA for x = 0.1,0.2,...,2.5.

(b) Compare your results with the LSA of the first lens.

(c) Do a ray tracing out to the paraxial image plane for the four lenses.

PROBLEM 61

Write a program to trace a set of parallel rays for the Tessar of Problem 22. Determine the LSA at the paraxial focal plane and compare it to the value at the circle of least confusion.

PROBLEM 62

The Tessar lens was first designed in 1902 and has had a reputation as a very good three-element lens for photography (the final doublet is regarded as a single element). An improvement has been proposed by K. D. Sharma, *Applied Optics* 19, 698 (1980), who altered the design by splitting the rear lens into two separate positive lenses, so that there are four separate elements. The specifications for the lens designed by Sharma are

c	t	n
0.038072		
	3.19	1.6103
–0.000115		
	9.66	1.0
–0.035993		
	0.70	1.7004
0.025811		
	3.80	1.0
–0.020683		
	6.26	1.6103
–0.043796		
	0.30	1.0
0.017481		
	4.84	1.6103
–0.014265		

Compare the quality of this lens with the Tessar in Problem 61 using both a ray tracing and a computation of LSA.

PROBLEM 63

A way of reducing spherical aberration in an astronomical reflecting telescope was proposed by D. D. Maksutov, *Journal of the Optical Society of America* 34, 270 (1944). A *corrector plate* with the shape of a meniscus (Figure 2.9) is placed in front of the mirror and serves to alter the path of the marginal rays.

(a) Given a Maksutov system with the following parameters

$r_1 = -152.8$ $\qquad n = 1.5225$

$r_1 = -158.62$ $\qquad d_1 = 10.0$

$r_3 = -823.17$ $\quad d_2 = 539.1$

$D = 100.0$

write a program that will generate the ray tracing diagram of Figure 2.10. Put the image plane about 450 units to the left of the mirror. Remember that the index after reflection is negative and so is the translation.

(b) Verify the LSA calculation of Figure 2.11.

It is interesting to see how the meniscus corrector plate shifts the rays away from the axis, with this shift increasing toward the margin. The result is a substantial improvement in the sharpness of focus.

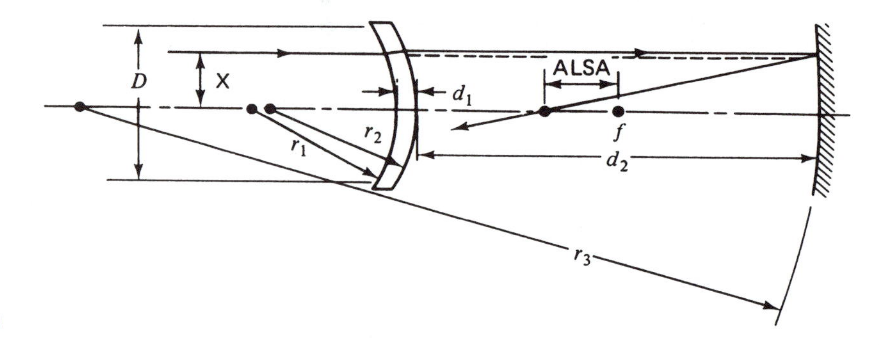

Fig. 2.9

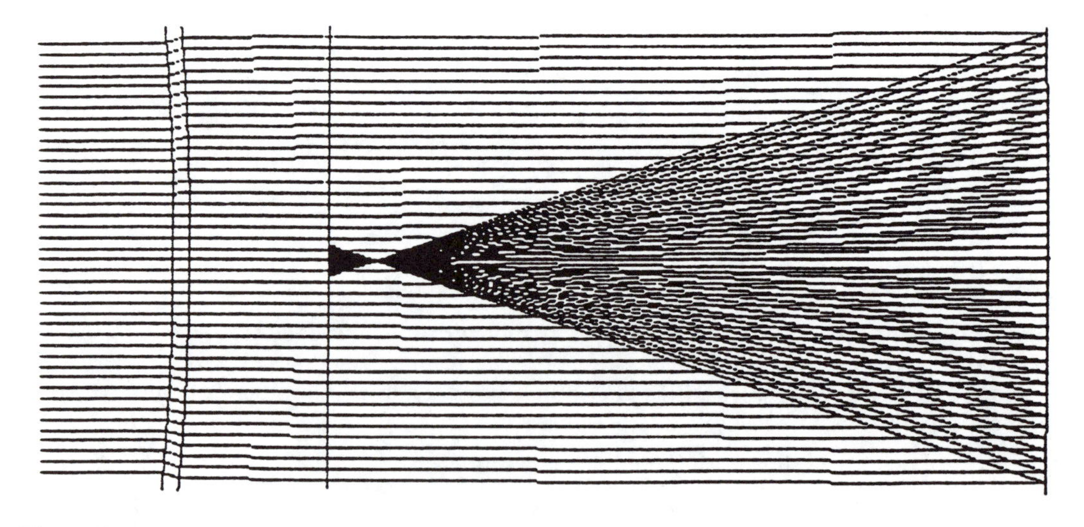

Fig. 2.10

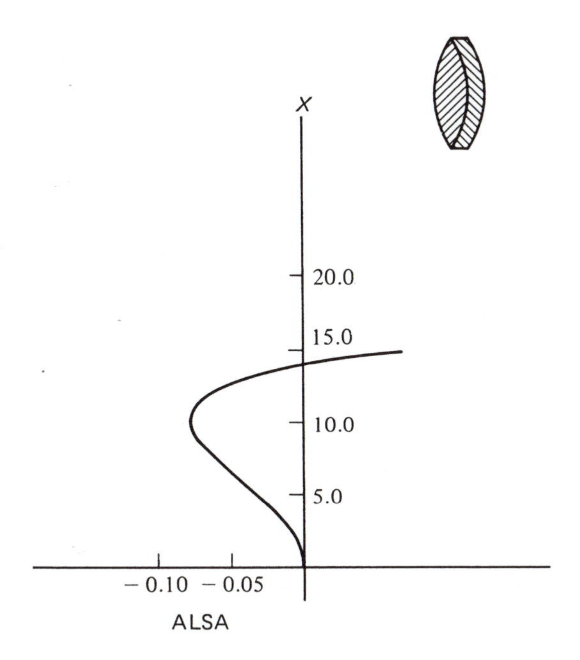

Fig. 2.11

PROBLEM 64

Figure 2.12 shows the ray tracing for a double concave lens. Write the program to generate this diagram. The BASIC procedure for a blue (color 9) dotted line joining (z_1, x_1) and (z_2, x_2) is

```
LINE (Z1,X1)-(Z2,X2), 9, , &HFFOO
```

That is, it requires a hexadecimal specification.

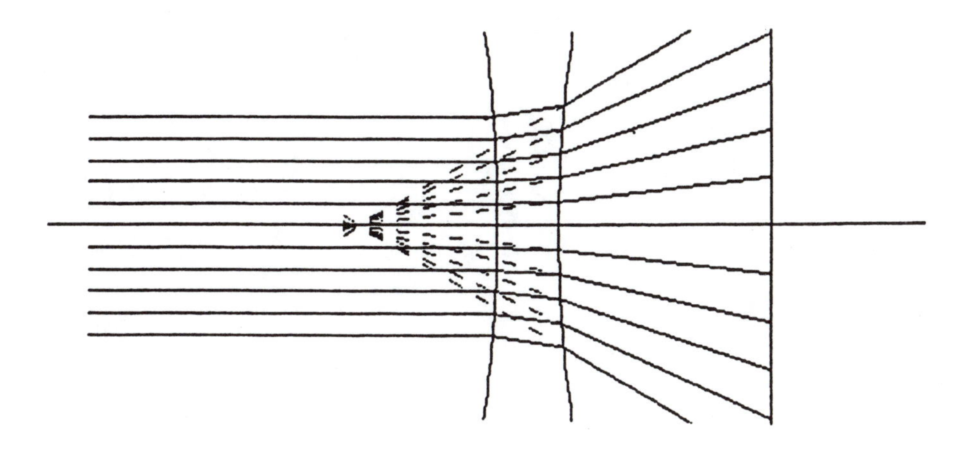

Fig. 2.12

CHAPTER 3

Nonparaxial Skew Optics

3.1 Skew Ray Tracing

We have seen how to extend matrix methods to nonparaxial, meridional rays and found that spherical aberration is due to the fact that the intermediate and marginal rays do not behave like the central rays. We now wish to deal with rays that do not lie in a meridional plane; these are *nonmeridional* or *skew* rays. Calculating the behavior of such a ray would appear to be a complicated problem, but this process turns out to be a simple extension of what we already know how to do. Let a skew ray start at an arbitrary point in space with coordinates x, y, and z and let it have direction-cosines L, M, and N with respect to OX, OY, and OZ. The points P and P_1 in Figure 2.1 move out of the plane ZOX, and, when the vectors in this figure are expressed as the sum of three, rather than two, components, equation (117) becomes

$$R_1^2 = R^2 + T_1^2 + 2T_1(zN + xL + yM) \tag{153}$$

The result is that a term relating to the y-axis has been added. This occurs also in the equation for surface 1, which is now a sphere rather than a circle, and equations for E and F become

$$\mathrm{E} = c_1[(z - v_1)\mathrm{N} + x\mathrm{L} + y\mathrm{M}] - \mathrm{N}$$

and

$$F = c_1[(z - v_1)^2 + x^2 + y^2] - 2(z - v_1)$$

We also have a third matrix equation, and this will be of the same form as (135). The additional equation corresponding to (133) that goes into the program will be

$$y_1 = y + T_1 M$$

The same thing happens for the refraction process. There is an additional matrix equation like (147) involving the old and new direction-cosines, and (141) becomes

$$\cos\theta_1 = N_1(z_1 - r_1)/r_1 - L_1 x_1/r_1 - M_1 y_1/r_1$$

Returning to the beginning of the previous chapter and using the lens considered there, let's trace a ray starting at $z = 0$, $x = 20$, and $y = 0$, with direction-cosines $L = -0.1$ and $M = 0.1$ (N is automatically determined since the sum of the squares of the direction-cosines is unity). The program steps that have to be changed start with the specification of the ray parameters as

```
X = 20: Y = 0: Z = 0
EL = -.1: EM = .1: EN = SQR(1 - EL * EL - EM * EM)
```

In the translation calculation, the new steps are

```
E = C(J) * (X * EL + Y * EM + ZV * EN) - EN
F = C(J) * (X * X + Y * Y + ZV * ZV) - 2 * ZV
```

There is a term for Y

```
Y + Y + SD * EM
```

And in the refraction portion, the new terms are

```
COSTHETA = EN * OP - C(J) * (EL * X + EM * Y)
```

and

```
EM = (N(J) * EM - K * Y)/ NP(J)
```

Adding also the printing commands for X, Y, Z and EL, EM, EN, we obtain the numerical results given in Table 3.1.

Table 3.1 Output of Skew Ray Tracing Program

x, L	*y, M*	*z, N*
–0.654624	20.654623	4.470334
–0.061857	–0.085074	0.994452
–1.046035	20.116308	–4.237134
–0.079774	–0.377840	0.922427
–7.078610	–8.456162	0.000001
–0.079774	–0.377840	0.922427

It can be seen that a ray leaving a point 20 units from the origin on the x-axis and pointing downward with respect to this axis but outward with respect to the y-axis will cross the paraxial focal plane at a point in the third quadrant of this plane.

3.2 COMA

An aberration due to skew rays called *coma* is difficult to explain and to understand. H. G. Zimmer in his book *Geometrical Optics* (Springer [1970]) says that there is no exact theory of coma. We shall show how a numerical approach gives an understanding, as well as corrects some misconceptions in the literature. Figure 3.1 shows a set of rays on a cylinder that is centered about the z-axis of a lens, so that all the rays are meridional. These rays are fairly far from the axis, so that they are also nonparaxial, but they all meet at a common image point, forming a cone whose apex is this image point. Now give this cylinder a downward displacement while holding fixed the intersection point of each ray with the dotted circle on the front of the lens. This tilting of the cylinder changes all its rays—except the top and the bottom rays—from meridional to skew. In addition, the cone on the image side will then tilt upward; it should not change in any other way if the skew rays continue to meet at a well-defined apex. But this is not what happens in a simple lens, as I will now explain.

Let the top and bottom meridional rays meet at a point P in image space and use this point to determine the location of an image plane (Figure 3.2). The skew ray just below the uppermost ray in Figure 3.1 will pass through this plane fairly close to the intersection point. Let's guess (and confirm later) that it will pass through a point

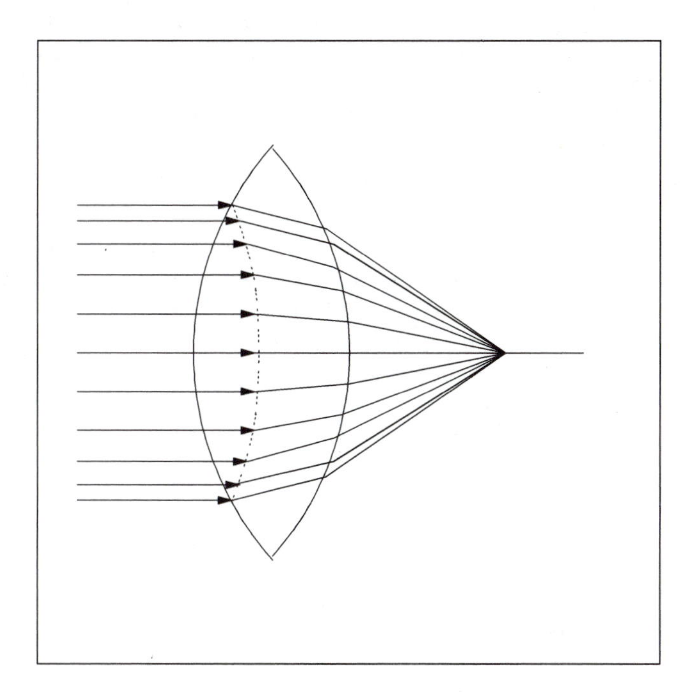

Fig. 3.1

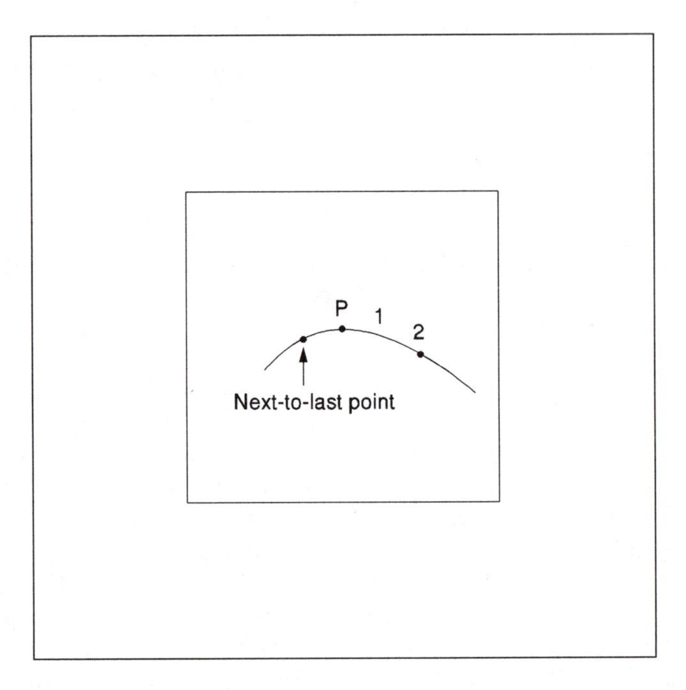

Fig. 3.2

slightly to the right of and a little below P, as we observe this plane from the lens position. This is point 1 in Figure 3.2. The next skew ray on the front of the lens will strike the image plane still farther below and to the right, producing point 2 in the figure. As we continue to trace these rays down the front half of the circle, we realize that a ray almost at the bottom of Figure 3.1 will have to be very close to the lower meridional ray and that its image point is just below but *to the left* of P; this is the point labeled "next-to-last" in Figure 3.2. All of these points join to form a closed curve. The back halves of the dotted circle and of the ray cylinder are not shown in the figure, but, since we are again starting at P in Figure 3.2, the same pattern will repeat itself. That is, we get a second closed curve that—by symmetry—should be identical to the first curve. This is a most unusual effect; double-valued phenomena are quite rare. To confirm this unusual prediction and to see what the double closed curve looks like, we use the skew ray tracing program to work out Problem 65. Part (a) of Problem 65 shows how to specify the distance of the intersection point of the two meridional rays from the second vertex of the lens, thus locating the image plane. Part (b) shows how to calculate the necessary slope information. The remainder of the problem deals with the way in which the skew rays are traced and the appearance of the coma pattern on the computer screen.

PROBLEM 65

(a) Tilt the cylinder of rays in Figure 3.1 as described above. Let the two meridional rays at the top and bottom, respectively, leave surface 2 of the lens and meet at the point P of Figure 3.2. Show that the distance from the vertex to the plane determined by their intersection is

$$t_2' = [(x_{22} - x_{21}) - m_2(z_{22} - z_{21})]/(m_1 - m_2)$$

where z_{21}, x_{21}, and m_1 are the coordinates and slope of the first ray as it emerges from surface 2, where z_{22}, x_{22}, and m_2 apply to the second ray, and—as before—the values along the axis are measured from V_2.

(b) Show that the slope m of each ray is the ratio L/N for that ray.

(c) Consider a double convex lens with radii of 10 and –10, thickness of 2, and index of 1.5. Find the intersection plane for a pair of parallel rays with direction-cosines $L = 0.01$ and $M = 0$ striking the first surface at a distance of +1 or –1 from the z-axis.

(d) Trace a group of equally spaced rays that start at the circle of unit radius on the face of the lens. Show the double closed curve produced on the image plane.

(e) Repeat for circles of radii 0.9, 0.8, ..., 0.3.

The graphical output of this program is shown in Figure 3.3. Each half-circle on the front side of the lens produces a closed figure (an approximation to a circle) as an image. If coma were the only aberration (when it is called *ideal coma*), the plots would be perfect, coinciding with the circles of Figure 3.4. The comet-like shape explains where the name comes from. We can picture the generation of these circles in the way

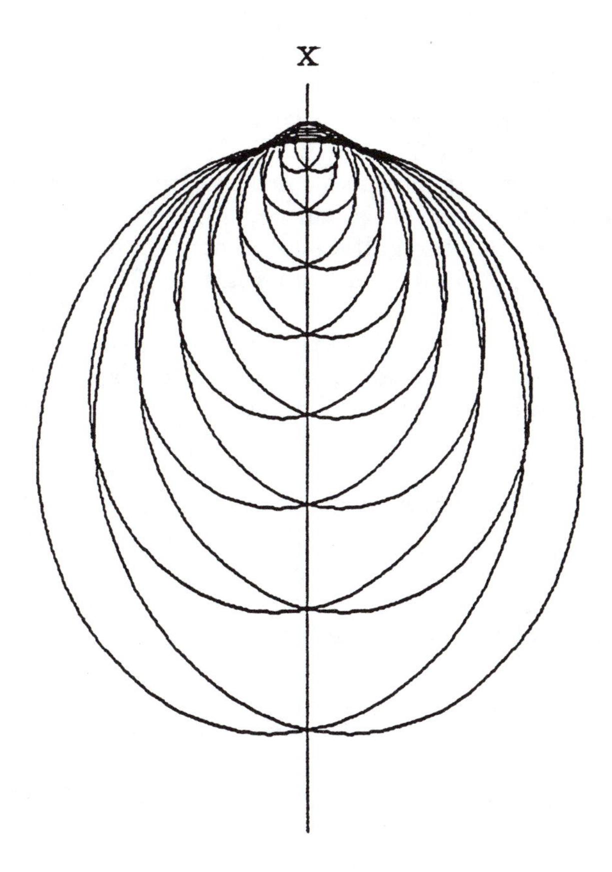

Fig. 3.3

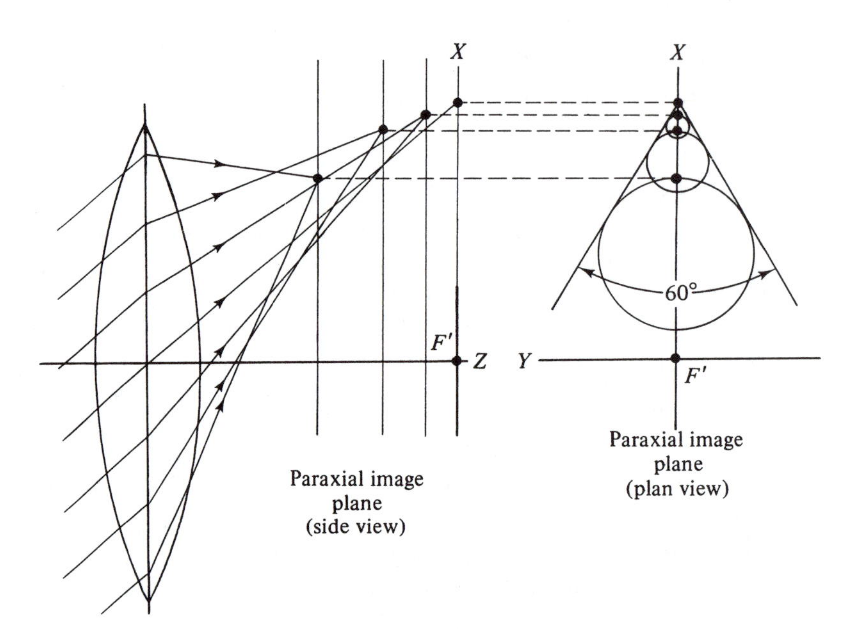

Fig. 3.4

shown in this figure. The ray through the geometrical center of the lens produces a point in the paraxial focal plane. The top and bottom rays (the meridional rays) from the circles of increasing size produce points that define image planes that are successively closer to the lens; this is a consequence of spherical aberration. Moving the top ray parallel to itself and along the half-circle generates the image plane pattern. The lines tangential to the coma circles make an angle of 60° with each other, as will be demonstrated in an analytic way later. The double curves of Figure 3.3, which are slightly offset to either side of the axis, exist because the simple lens used to generate Figure 3.3 has aberrations other than coma. Most optics books look at coma in a way different from that given here. Again referring to W. Smith, *Modern Optical Engineering*, 2d ed., McGraw-Hill (1990), coma is defined as the variation of magnification with aperture, analogous to the definition of spherical aberration as the variation of focal length with aperture. And as I previously mentioned, focal length and magnification are purely paraxial concepts, not applicable to the description of aberrations.

The definition that comes from the arguments given earlier is that *coma is the failure of skew rays to match the behavior of meridional rays*, just as spherical aberration is the failure of meridional rays to match the behavior of paraxial rays. Most texts show coma in its ideal form. One of the few sources presenting illustrations like Figure 3.3 is G. Franke, *Physical Optics in Photography*, The Focal Press (1966). It is interesting to observe that these double curves show up in photographic images as indicated in Figure 3.5; the two overlapping circles form a *cardiod* or heart-shaped pattern. This effect has been demonstrated by B. Sherman, *Modern Photography* 32, 118 (1968).

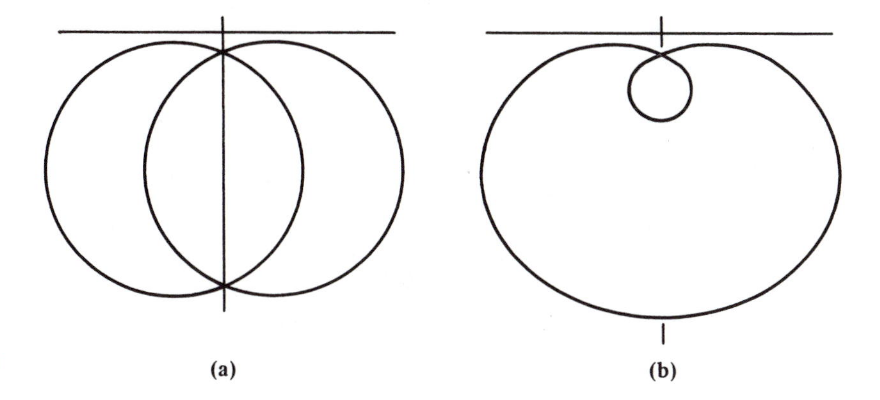

Fig. 3.5

PROBLEM 66

The optimal shape lens discussed in Section 2.3 is one for which the spherical aberration has been minimized. It has been found that coma is also reduced for such lenses. Compare the LSA and the coma for a double convex lens with radii of 50 and –50, thickness of 15, and index of 1.5 using shape factors of –2.00, –1.00, –0.50, 0, 0.50, 1.00, and 2.00.

PROBLEM 67

Repeat the coma calculations of Problem 66 by considering a point object on an axis (for which you know the coma is zero), and have the object move away in reasonable-sized steps until the coma becomes very noticeable.

3.3 ASTIGMATISM

We have shown that spherical aberration is the failure of meridional rays to obey the paraxial approximation and that coma is the failure of skew rays to match the behavior of meridional rays. To see the connection between these two aberrations, consider the object point P of Figure 3.6. This very complicated diagram can actually be very helpful in understanding the third of the five geometrical aberrations. Let P be the source of a meridional fan: this is the group of rays with the top ray labeled PA and the bottom ray labeled PB. If the lens is completely corrected for spherical aberration, this fan will have a sharp image point P_T' lying directly below the z-axis. Now let P also be the source of a fan bounded by the rays PC and PD. This fan is at right angles to the other one; we can think of these rays as having the maximum skewness possible. If the lens has no coma, this fan also will produce a sharp image point P_S' which, for a converging lens, will be farther from the lens than the tangential focal point. In other words, even though the lens has been fully corrected for both aberrations, the two corrections will not necessarily produce a common image. In the figure, the meridional fan is called a *tangential* fan and the skew fan is called a *sagittal* fan; these are the terms commonly used by lens designers. The failure of the sagittal and tangential

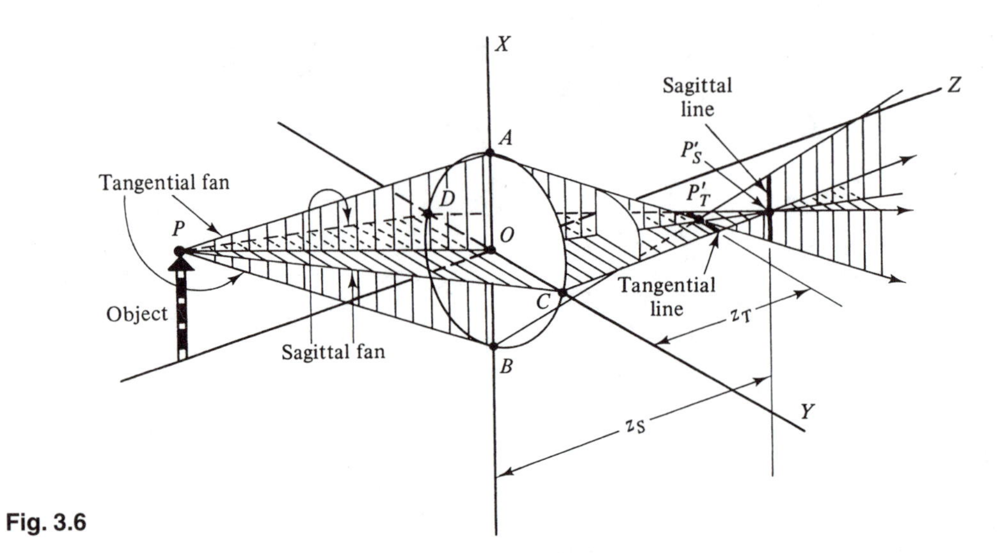

Fig. 3.6

rays to produce a single image in a lens corrected for both spherical aberration and coma is known as *astigmatism*.

Astigmatism, coma, and spherical aberration are the *point aberrations* that an optical system can have. We have already noted that spherical aberration is the only one that can be associated with a point on the z-axis; the others require an off-axis object. Astigmatism is very common in the human eye; to see how it shows up, look at the sagittal image point P_S' of Figure 3.6. This point is a distance z_S from the x,y-plane. A plane through this point will have a line image on it, the *sagittal line*, due to the tangential fan, which has already come to a focus and is now diverging. The converse effect, producing a *tangential line*, occurs at the other image plane. If we locate an image plane halfway between the two, then both fans contribute to the image, and in the ideal case it will be a circle. As the image plane is moved forward or backward, these images become ellipses and eventually reduce to a sagittal or a tangential line, as shown in Figure 3.7.

Because there are two different image planes, an object with spokes (Figure 3.8a) will have an image for which the vertical lines are sharp and the others get gradually poorer as they become more horizontal, or vice versa, depending on which image plane is chosen (Figure 3.8b). Eye examinations detect astigmatism if the spokes appear to go from black to gray. This example of the effect of astigmatism was often based on a radial pattern of arrows, which explains the origin of the word sagittal, derived from the Latin *sagitta* for arrow.

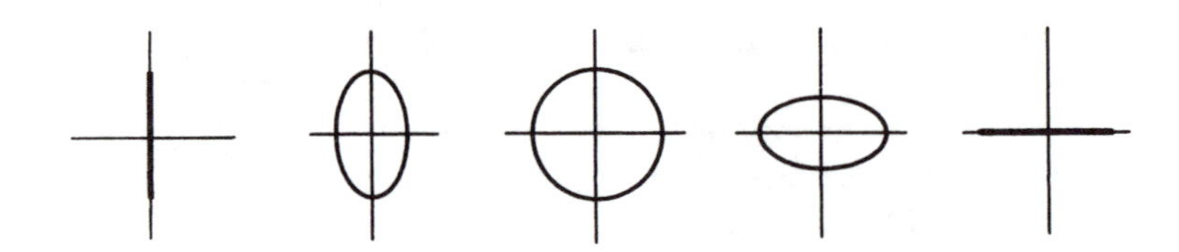

Fig. 3.7

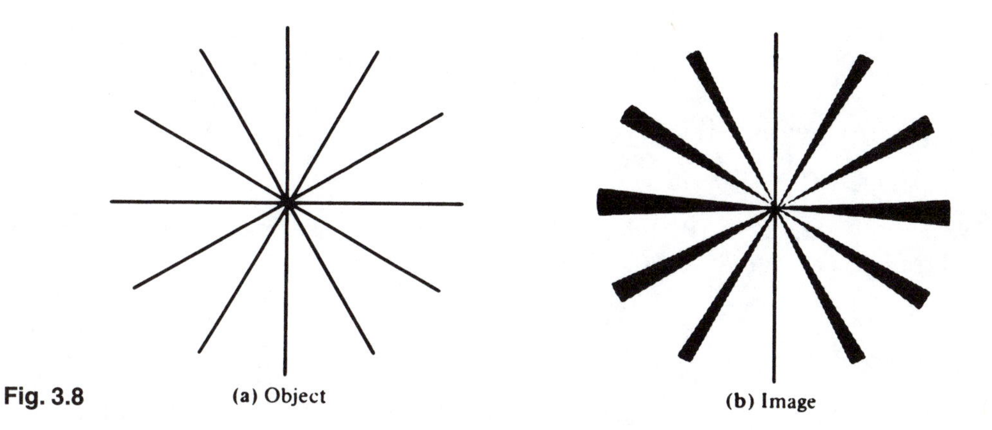

Fig. 3.8 (a) Object (b) Image

3.4 APLANATIC ELEMENTS

Figure 3.9 shows an unusual situation: an object PQ lies *inside* a glass sphere of radius r and index n; the sphere is immersed in a medium of index n'. To find the image Q′ of Q (a virtual image), trace two rays. One will lie on the axis and be undeviated. The other will be refracted at the right-hand surface in such a way that θ' is greater than θ for $n' < n$, and its extension meets the axis at Q′. By the law of sines for oblique triangles, we have

$$QC/\sin\theta = r/\sin\alpha,\ Q'C/\sin\theta' = r/\sin\alpha'$$

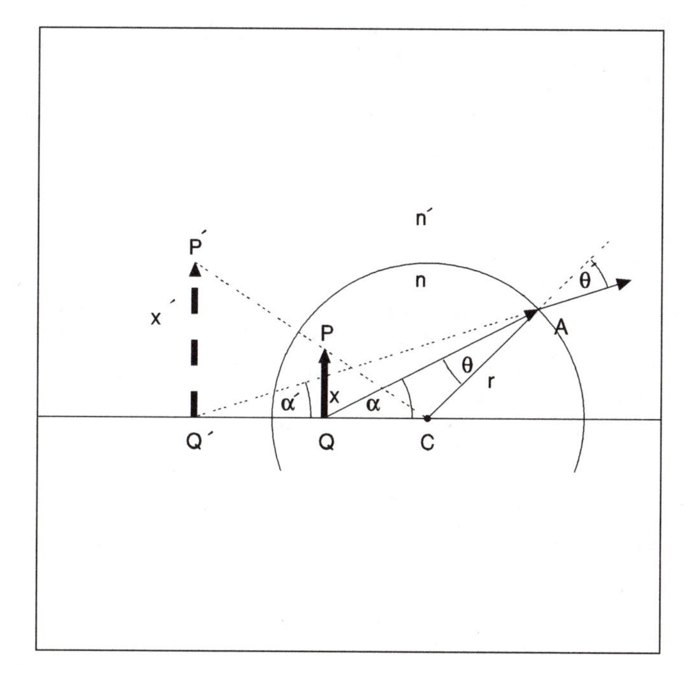

Fig. 3.9

and by Snell's law

$$n \sin \theta = n' \sin \theta'$$

Combining these three equations gives

$$QC/QC' = (\sin \theta/\sin \alpha)(\sin \alpha'/\sin \theta') = (n' \sin \alpha')/(n \sin \alpha)$$

Now consider the point P; its image at P′ is found by tracing an undeviated ray backward from the center C. Then

$$QC/x = Q'C/x'$$

and these last two equations combine to give

$$nx \sin \alpha = n'x' \sin \alpha'$$

An important special case is specified by making QC equal to rn'/n. Then

$$rn'/n = r \sin \theta/\sin \alpha$$

It follows from Snell's law that $\sin \alpha = \sin \theta'$. Figure 3.9 shows that $\alpha = \alpha' + \theta' - \theta$, from which $\alpha' = \theta$ and again from Snell's law

$$Q'C = rn/n'$$

We have thus located the object and image in such a way that their positions are independent of the ray angles α and α'; all rays leaving Q will meet at Q′, and there is no spherical aberration. If the object PQ is small, we can predict that the rays from P will also form a sharp image. That is, there will be no coma due to the skew rays coming from P. Such a system is said to be *aplanatic*. This analysis helps us to understand the *oil immersion microscope objective*. Grind away a section of the glass sphere of Figure 3.9, leaving a flat surface normal to the axis at Q, and place an object in the center. There will be an aberration-free image at Q′. This component, called a *hyper-hemisphere*, is placed close to the very thin cover glass of a microscope slide, shown (not to scale) in Figure 3.10a.

PROBLEM 68

(a) Consider first the left-hand half of Figure 3.10a. Taking the index of refraction of the glass slide as 1.6, use the concept of total internal reflection (Section 2.4) to show that the maximum angle that rays leaving the point object and the slide can have is about 40°. Then show that a layer of cedarwood oil (n = 1.517), as on the right, raises this angle to about 70°. Thus, the oil immersion objective combines high image quality and great resolution due to the wide field of view.

(b) Prove by ray tracing that the relations between α,θ' and α',θ are valid for $\sin \alpha = 0.8$.

(c) Using the system matrix for a single refracting surface, show that the positions of object and image in Figure 3.9 are paraxially correct.

(d) Let r = 10 and n = 1.5. Place a small object at a distance r/n from C, locate the cardinal points, and show the image on a scale diagram.

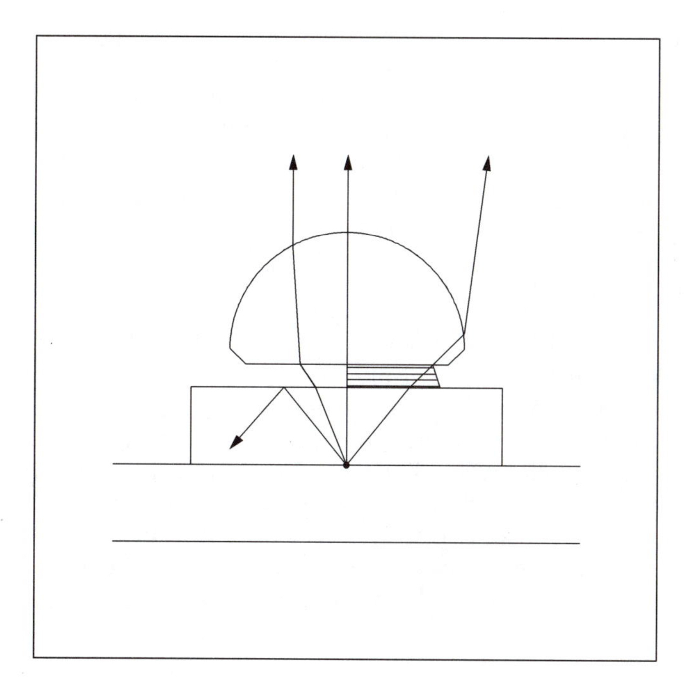

Fig. 3.10a

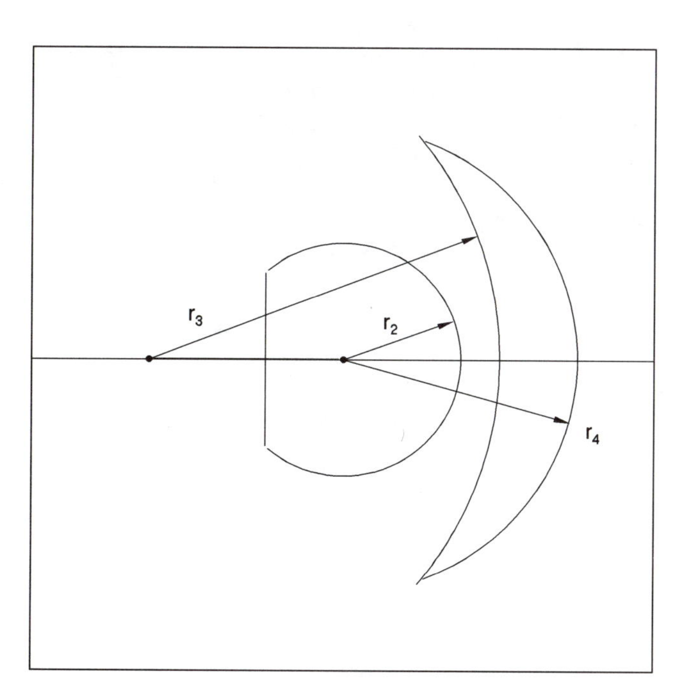

Fig. 3.10b

(e) An additional way to take advantage of the aplanatic behavior is to add a second, meniscus-type lens, shown in Figure 3.10b and specified as follows: its front surface is centered at the image point Q′ and its rear surface is located so that it produces an aplanatic image of the rays coming from Q′. Show that the magnification of the hyper-hemisphere is equal to n^2 and that the magnification of the meniscus lens is n.

(f) Let the index of both lenses in Figure 3.10b be 1.5. Let r_2 be 108 units and r_3 be 290 units. Find r_4 and write a program to locate the final image and to show that the combined magnification is n^3.

The aplanatic pair of lenses shown in Figure 3.10b behaves like a paraxial system, since it is free of aberrations. For such a system, the magnification is a constant regardless of what rays produce the image, so that $n' \sin \alpha' / n \sin \alpha$ is constant. This result is known as the *Abbe sine condition*; it is used by lens designers as a measure of the amount of coma. The violation of the Abbe condition is called *the offense against the sine condition (OSC)*. It is defined in such a way that only a few rays need be traced in order to apply it—and this was quite an important consideration in the days when optical computations were done by hand, using logarithms. Note that this condition requires tracing only meridional rays, which made it easier to apply. Now that we have programs to calculate the size of the coma pattern, we have a much better way of studying this aberration in detail.

3.5 Curvature of Field and Distortion

Having covered the three point aberrations, the two remaining lens defects are associated with extended objects. If we move the object point P in Figure 3.6 closer to or farther from the z-axis, we would expect the positions of the tangential and sagittal focal planes to shift, for it is only when the paraxial approximation holds that these image points are independent of x. Hence, we obtain the two curves of Figure 3.11, which shows what astigmatism does to the image of a two-dimensional object. If the astigmatism could be eliminated, the effect would be to make these curved image planes

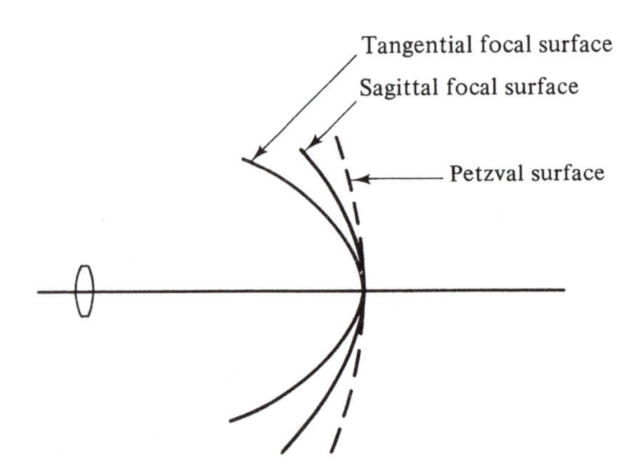

Fig. 3.11

coincide, but we have no guarantee that the common plane will be flat, or paraxial. The resulting defect is called *Petzval curvature* or *curvature of field*. For a single lens, the Petzval surface can be flattened by a stop in the proper place, and this is usually done in inexpensive cameras. Petzval curvature is associated with the z-axis. If we take the object in Figure 3.6 and move it along the y-axis, then all rays leaving it are skew; this introduces *distortion*, the aberration associated with the coordinates normal to the symmetry axis. Distortion is what causes vertical lines to bulge outward (*barrel distortion*) or inward (*pincushion distortion*) at the image plane, as shown in Figure 3.12.

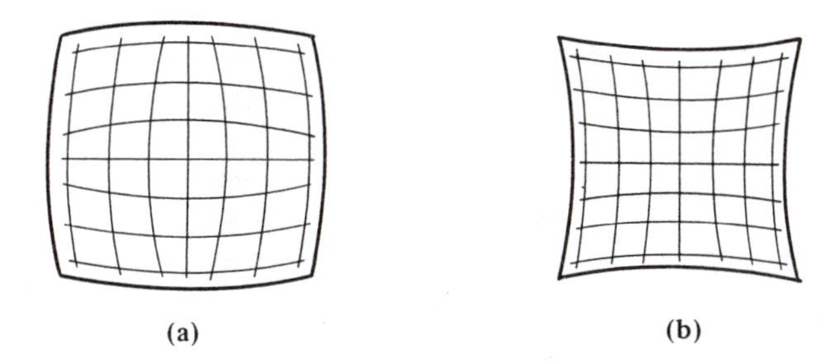

Fig. 3.12 (a) (b)

A pinhole camera will have no distortion since there are no skew rays, and a single thin lens with a small aperture will have very little. Placing a stop near the lens to reduce astigmatism and curvature of field introduces distortion because, as shown in Figure 3.13a, the rays for object points far from the axis are limited to off-center sections of the lens. The situation in this figure corresponds to barrel distortion; placing the stop on the other side of the lens produces pincushioning. Distortion can therefore be reduced by a design that consists of two identical groupings (Figure 3.13b) with the iris diaphragm placed between them; the distortion of the front half cancels that of the back half. The percentage of distortion can be quantitatively expressed as $D = 100\,(I' - I)/I$ where I is the paraxial image size and I' is the size of the distorted image.

PROBLEM 69

Astigmatism and field curvature can be treated by an extension of the method used for coma. Start with a tipped cylinder that strikes the front face of a lens on a circle of radius R. Let the axis of the cylinder be OZ. The rays that strike the lens at $x = \pm R$, $y = 0$ are tangential, and the two rays that strike it at $x = 0$, $y = \pm R$ are sagittal. Find the intersection points of each pair in image space. The z-coordinate of these two image points with respect to the paraxial image plane is a measure of the tangential and sagittal field curvatures, respectively, while their difference measures the astigmatism. Derive formulas for each of the two intersection points and use them to compute and plot the distance from the tangential and sagittal focal planes and the paraxial image plane as a function of image height for the lens whose radii, spacings, and indices are

$$r = 40.94,\ \infty,\ -55.65,\ 39.75,\ 107.56,\ -43.33$$

$$t' = 8.74,\ 11.05,\ 2.78,\ 7.63,\ 9.54$$

$$n' = 1.617,\ 1.649,\ 1.617$$

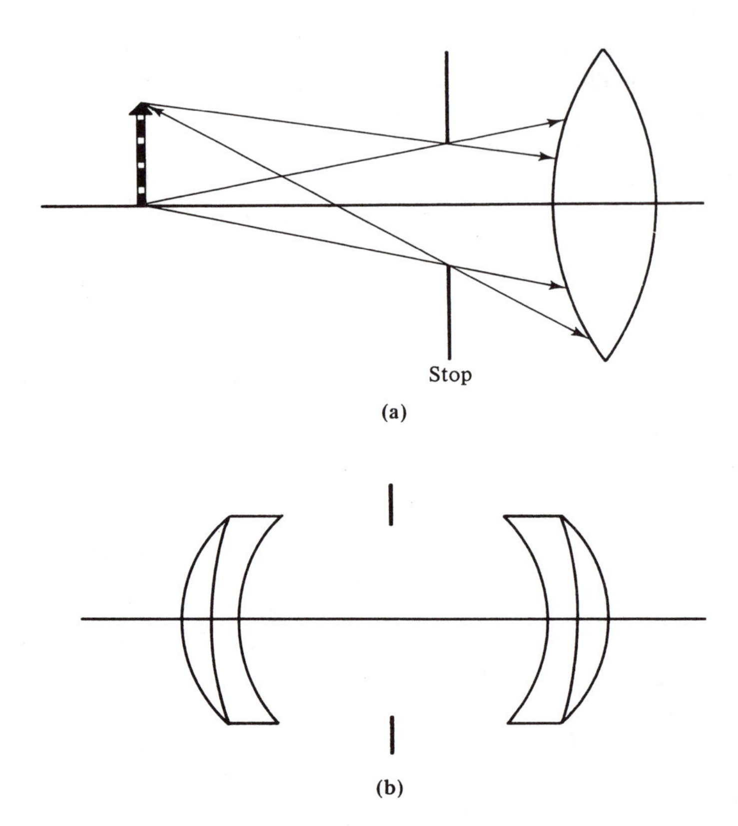

Fig. 3.13

Figure 3.14 shows the output for a calculation using approximate formulas rather than exact ray tracing. Note that the designer has arranged for astigmatism to vanish at the lens margin.

This example indicates the nature of astigmatism in a system for which the coma and the spherical aberration are low but not necessarily zero; it is the failure of the tangential and sagittal image planes to coincide.

PROBLEM 70

Problem 60 shows the design of a lens to give low spherical aberration. This idea will work, however, only if the lenses are very accurately aligned. Even a small error in assembly will mean that meridional rays become skew and coma is introduced. It has been proposed that a fifth element, shown in Figure 3.15, will reduce coma. The design is as follows:

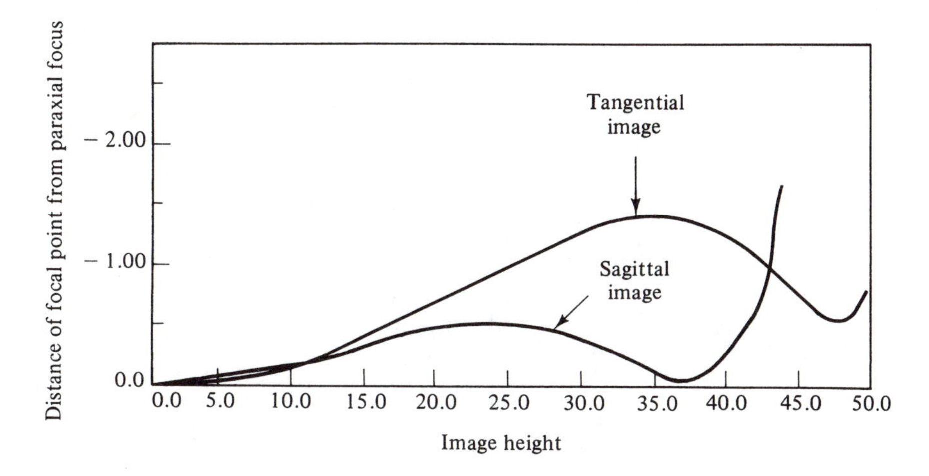

Fig. 3.14

r	t'	n'
117.20		
	1.35	1.5536
–2020.0		
	0.00	
–60.25		
	1.35	1.5536
132.10		
	0.00	
39.50		
	1.45	1.5536
61.50		
	0.00	
29.15		
	1.60	1.5536
38.68		
	0.00	
16.55		
	3.50	1.7190
22.75		

(a) Using a direction-cosine with respect to the axis of 0.019 and letting $x = 3.7$, show that the coma is minimized when the first lens is changed such that

$$r_1 = 113.05,\ r_2 = -2010.5$$

(b) Show that another way to minimize the coma is to change t_8' to a value of about 1.0. Find this value.

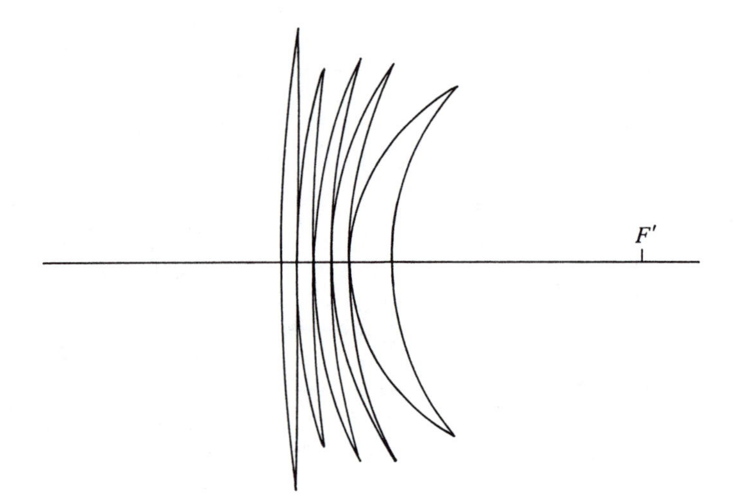

Fig. 3.15

CHAPTER 4

Conic Aspheric Surfaces

4.1 Nonspherical Surfaces

So far, we have assumed that optical systems use only spherical or plane reflecting and refracting surfaces. However, many astronomical telescopes use parabolic and hyperbolic mirrors, and there was also a high-speed camera lens (the Leitz Noctilux) for which one surface was not spherical. In recent years, nonspherical or *aspheric* glass lenses have become more common because they can be mass produced by computer-controlled machinery, and plastic lenses can be molded with very strict tolerances. We shall see how ray tracing procedures can easily be extended to the simplest kind of aspheric surfaces: the conic sections, using skew ray matrix methods and the numerical calculations previously given.

4.2 The Geometry of Conic Sections

Let the point P with coordinates (x,y) in Figure 4.1 be allowed to move and let the point F on the x-axis be stationary. If the distance from P to F (the *focus*) and the perpendicular distance to the y-axis maintain a constant ratio e as P moves, the resulting curve or surface is a *conic section*, and e is its *eccentricity*. This definition is expressed as

$$e = \frac{\overline{FP}}{\overline{DP}} = \frac{\overline{FP}}{\overline{OM}} = \frac{\sqrt{(x - p)^2 + y^2}}{x} \tag{154}$$

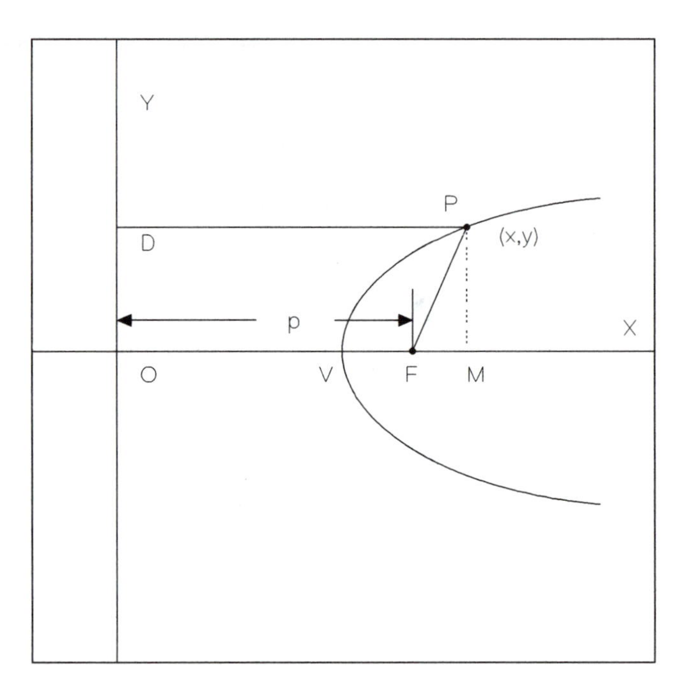

Fig. 4.1

where p is the x coordinate of the focus. Planar conic sections have names determined by the value of the eccentricity. These are

$e < 1$ ellipse ($e = 0$ for a circle)

$e = 1$ parabola

$e > 1$ hyperbola

Equation (154) for the parabola becomes

$$y^2 = 2xp - p^2$$

If we replace x with $x + (p/2)$ in this equation, the origin will be shifted to the vertex V and the equation is simplified to

$$y^2 = 2px = 4fx \tag{155}$$

where $f = p/2$ is the location of F (Figure 4.2). This well-known result defines a parabola for which the y-axis in its new position is tangential to the vertex. I will show below that the geometrical focus F will also be the optical focus of a parabolic mirror with focal length f. Equally well known are the equations for the ellipse

$$\frac{x^2}{a^2} + \frac{y^2}{b^2} = 1 \tag{156}$$

and the hyperbola

$$\frac{x^2}{a^2} - \frac{y^2}{b^2} = 1 \tag{157}$$

but these relations define figures that are symmetrical about both axes, and it is convenient for ray tracing purposes to have the vertices located as shown in Figure 4.2. It is also convenient to have a single defining equation for any kind of conic section, rather than the three separate equations (155), (156), and (157).

PROBLEM 71

It has been known for many years that a parabola will perfectly focus rays parallel to its axis. This can be demonstrated from Fermat's principle if we extend the statement given in Problem 2 (Chapter 1) to cover the possibility that the travel time is both a maximum or minimum; that is, the time on any path is constant. Consider a ray along the axis and another ray parallel to it (Figure 4.2). Show that the reflected rays will be equal in length if they meet at the focus, whose location is a distance $p/2$ or f from the y-axis.

We shall show that the desired equation, in our ZOX coordinate system, is

$$\frac{C(x^2 - Sz^2)}{2} - z = 0 \tag{158}$$

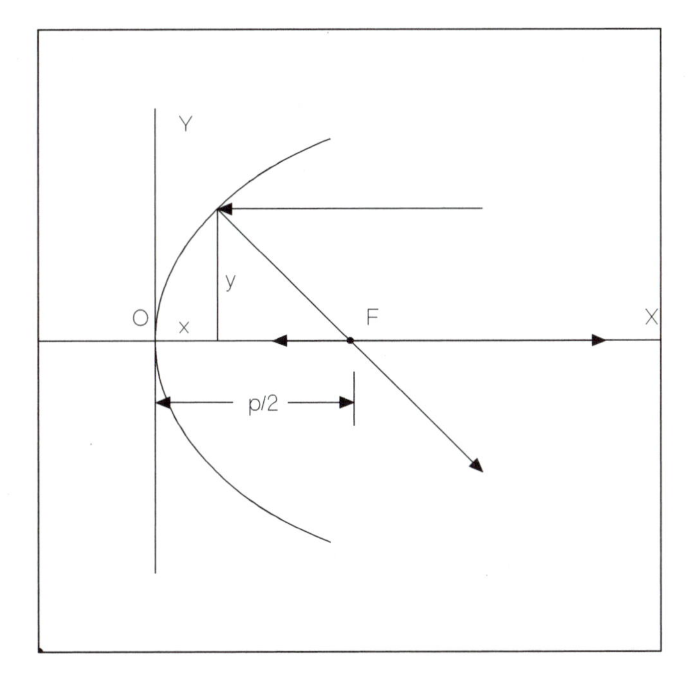

Fig. 4.2

The constant C is called the *vertex curvature*, while S is the *shape factor*; it should not be confused with the quantity σ used in equation (150). To understand the significance of these quantities, let $C = 0$. Then (158) becomes

$$z = 0$$

which describes the vertical axis in the ZOX system. This is what we would expect for a curve with zero curvature. Next, let $S = -1$ and $C = 1/r$, obtaining

$$z^2 - 2zr + x^2 = 0$$

which we recognize as the equation of a circle whose left edge is at the origin. Thus, C for a circle is the same as the constant curvature $c = 1/r$. Now let $S = 0$ and C be arbitrary. Then (158) reduces to

$$x^2 = 2z/C$$

which is a parabola, with $C = 1/p$ or $1/2f$. In the region very close to the origin, the parabola is almost a circle of curvature c, and we have $f = 1/2c$ or $f = r/2$, which is our earlier result for the paraxial focus of a sphere. Replacing the sphere whose constant curvature is c with a parabola whose vertex curvature is C gives a mirror of the same focal length and no spherical aberration for parallel rays. Having established that $S = -1$ produces a circle, we would expect other negative values to give ellipses and positive values to generate hyperbolas, since the division at $S = 0$ corresponds to a parabola. That is, the conic sections are specified by S in the following way.

$S < 0$ ellipse ($S = -1$ for a circle)

$S = 0$ parabola

$S > 0$ hyperbola

Figure 4.3 illustrates a set of conic sections with constant vertex curvature and shows how the different kinds of conic sections are produced as S varies. Since some designers specify elliptical and hyperbolic refracting or reflecting surfaces in terms of the eccentricity e, we need a way of relating this quantity to the shape factor S. To do this, we take equation (154) and shift the origin to the vertex in Figure 4.1.

PROBLEM 72

(a) Show that the y-axis has to be shifted a distance

$$\overline{OV} = \frac{p}{1 + e} \tag{159}$$

(b) Changing x to $x + \overline{OV}$, show that (154) in the ZOX coordinate system becomes

$$z^2(1 - e^2) - 2pze + x^2 = 0 \tag{160}$$

Comparing this with (162) written as

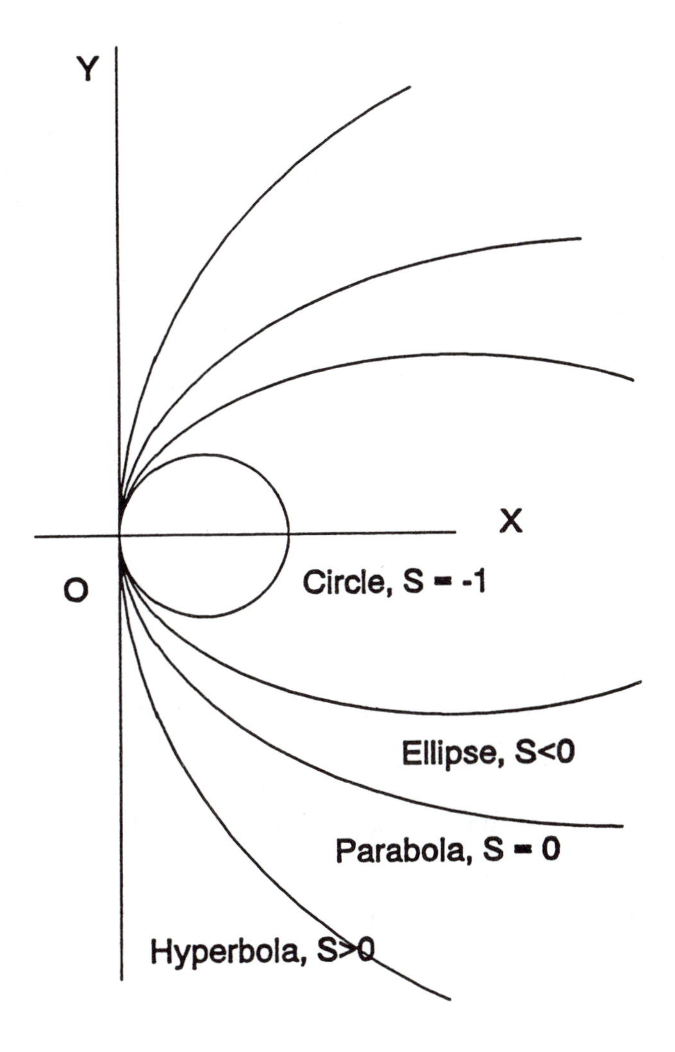

Fig. 4.3

$$-Sz^2 - 2\frac{z}{C} + x^2 = 0 \tag{161}$$

implies that

$$pe = \frac{1}{C} \qquad S = e^2 - 1 \tag{162}$$

which verifies the specification of a conic section by its shape factor S. This also agrees with the fact that $p = 1/C$ for a parabola, as previously obtained. We can use the results of Problem 72 to obtain some other useful information. If we let $y = 0$ in (160), we find that x has two possible values. These are

$$x = 0, \frac{2pe}{1-e^2} \tag{163}$$

The root with $1-e^2$ in the denominator is valid only when e is not equal to 1; that is, it applies to ellipses or hyperbolas. This shows that these two figures cross the axis twice and that half the distance between their vertices, the *semimajor axis*, is given by the formula

$$a = \frac{pe}{1 - e^2} \tag{164}$$

Another important quantity needed for ray tracing is the distance from vertex to focus expressed in terms of C and S.

PROBLEM 73

(a) Show that

$$\overline{VF} = \frac{\sqrt{1 + S} - 1}{CS} \tag{165}$$

in Figure 4.1.

(b) This expression gives the distance between vertex and focus for a conic section whose vertex does not touch the vertical axis. Show that it is still valid for the parabola of Figure 4.2.

The ellipse and the hyperbola, being symmetrical about both axes, will have two focal points that are equidistant from the center. Equation (165) stipulates the distance between the vertex and the nearer of the two foci. There are many applications for which we need to know the distance from a vertex to the other focus. We do this by adding (164) and (165) to obtain the result proven in Problem 79.

PROBLEM 74

(a) If a loop of string is placed around two pins and a pencil is held tightly against the loop, it is possible to draw an ellipse. Rotating an ellipse about the z-axis produces an ellipsoid of revolution. Put a reflecting coating on this ellipsoid and show that there is a connection between the way the ellipse is generated and Fermat's principle that demonstrates that a ray of light leaving one focus will be reflected in such a way as to pass through the other focus.

(b) Use this result to demonstrate that, if the source of light at the focus is a moderately powerful laser, a small hole will be burned through one vertex of the ellipsoid.

4.3 Ray Tracing for Conic Surfaces

We now extend the previous programs to conic sections. The translation process that was used is altered by replacing the equation of a sphere with the three-dimensional form of (158); this is

$$\frac{C_1}{2}[x_1^2 + y_1^2 - S_1(z_1 - v_1)^2] - (z_1 - v_1) = 0 \tag{166}$$

where v_1 is the location of the vertex. Combining this with (153), the resulting quadratic for the ray length is

$$T_1 = \frac{F}{-E + \sqrt{E^2 - DFC_1}} \tag{167}$$

where

$$D = 1 - (S_1 + 1)N^2 \tag{168}$$

$$E = C_1[xL + yM - S_1(z - v_1)N] - N \tag{169}$$

$$F = C_1[x^2 + y^2 - S_1(z - v_1)^2] - 2(z - v_1) \tag{170}$$

PROBLEM 75

Verify this algebra and show that these equations agree with previous expressions when the aspheric surface simplifies to a sphere.

The refraction process is also an extension of the previous procedure, but somewhat more complicated, because the unit normal is not as easy to specify. We use the fact that the gradient operator

$$\nabla = \hat{\boldsymbol{x}}\, \partial/\partial x + \hat{\boldsymbol{y}}\, \partial/\partial y + \hat{\boldsymbol{z}}\, \partial/\partial z \tag{171}$$

when applied to the defining equation of a surface will produce a vector that is normal to that surface. If the aspheric surface has its vertex at the origin, then by equation (157) it can be defined through the relation

$$F(x_1, y_1, z_1) = (C_1/2)(x_1^2 + y_1^2 - S_1 z_1^2) - z_1 \tag{172}$$

The components of the gradient are

$$\partial F/\partial x_1 = C_1 x_1, \quad \partial F/\partial y_1 = C_1 y_1, \quad \partial F/\partial z_1 = -C_1 S_1 z_1 - 1 \tag{173}$$

Dividing by the magnitude of the gradient, the unit normal is then

$$\boldsymbol{s}_1 = \frac{C_1 x_1 \hat{x} + C_1 y_1 \hat{y} - (C_1 S_1 z_1 + 1)\hat{z}}{\sqrt{C_1^2(x_1^2 + y_1^2) + (C_1 S_1 z_1 + 1)^2}} \tag{174}$$

PROBLEM 76

Show that the denominator of this expression reduces to unity for a sphere and that the numerator gives direction-cosines that were previously determined for spherical surfaces.

Using this result in (140), it becomes

$$\cos\theta_1 = \frac{-C_1x_1L_1 - C_1y_1M_1 + (C_1S_1z_1 + 1)N_1}{\sqrt{C_1^2(x_1^2 + y_1^2) + (C_1S_1z_1 + 1)^2}} \tag{175}$$

With this more involved expression for the unit normal, the three matrix refraction equations are now

$$L_1' = (n_1/n_1')L_1 - x_1K_1/n_1'\sqrt{C_1^2(x_1^2 + y_1^2) + (C_1S_1z_1 + 1)^2} \tag{176}$$

$$M_1' = (n_1/n_1')M_1 - y_1K_1/n_1'\sqrt{C_1^2(x_1^2 + y_1^2) + (C_1S_1z_1 + 1)^2} \tag{177}$$

$$N_1' = (n_1/n_1')N_1 - (C_1S_1z_1 + 1)K_1/n_1'\sqrt{C_1^2(x_1^2 + y_1^2) + (C_1S_1z_1 + 1)^2} \tag{178}$$

These equations will of course reduce to the set given in Chapter 3 for a three-dimensional spherical surface.

4.4 Parabolic Mirrors

As our first example of ray tracing for optical systems with aspheric refraction or reflection surfaces, we consider a very useful device involving two identical parabolic mirrors. An arc lamp is a good source of heat but, unfortunately, there is a large amount of pollution from the disintegrating electrodes. To melt a sample in a clean atmosphere, we place the lamp at the focus of a parabolic mirror, let the emerging parallel rays travel across the room, and concentrate them with a second mirror. For the example of Figure 4.4, the vertex curvature is 100 units, so that the foci will be at 50 units, and the spacing chosen is 250 units.

The following constants are used:

Vertex curvatures C	–0.01, 0.01, 0
Incident indices n	1, –1, 1
Refracted indices n'	–1, 1, 1
Shape factors S	0, 0, 0
Translations T	50, –200, 50

The values of C apply to the two mirrors and the image plane. The index n is for air and it reverses sign after each of the two reflections. The shape factor S is 0 for a

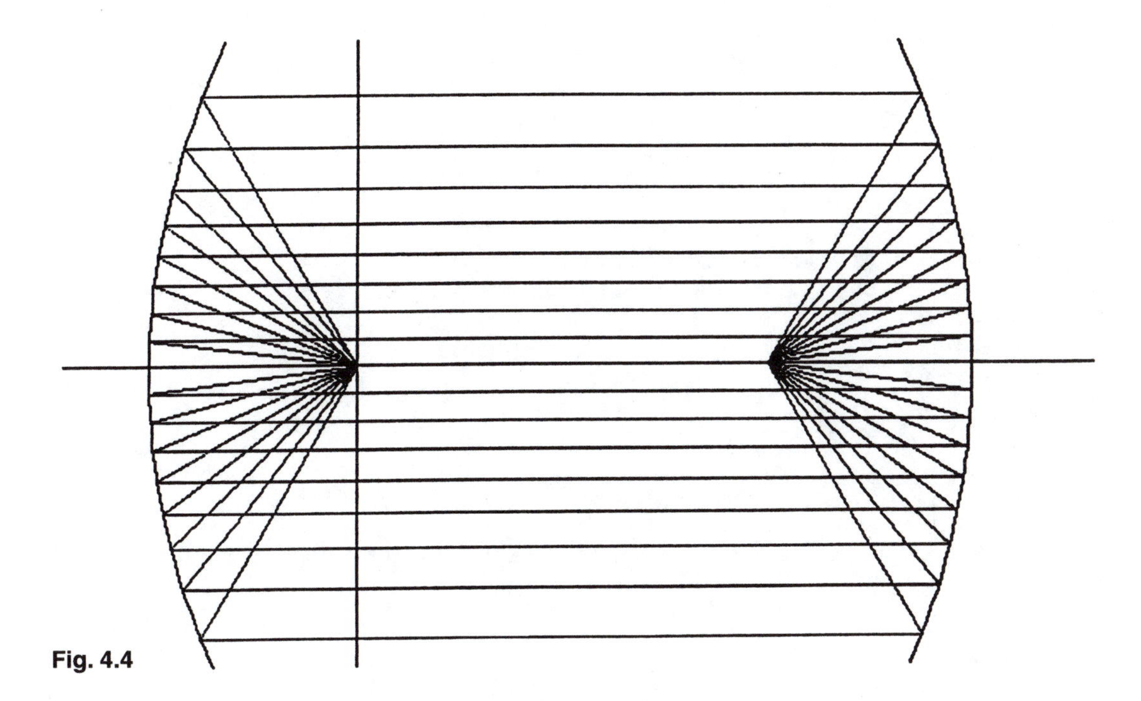

Fig. 4.4

parabola and arbitrary for the image plane. The translation T reverses sign twice just like n. All of these constants are in the program at the beginning.

The program using the translation and refraction procedures just developed is

```
CLS : SCREEN 9
WL = -180: WR = 80: WD = -60: WU = -WD
WINDOW (WL, WD)-(WR, WU)
T(1) = 50: T(2) = -200: T(3) = 50
C(1) = -.01: C(2) = .01: C(3) = 0
N(1) = 1: N(2) = -1: N(3) = 1
NP(1) = -1: NP(2) = 1: NP(3) = 1
S(1) = 0: S(2) = 0: S(3) = 0
EL0 = -.8
10 X = 0: Y = 0: Z = 0
Z1 = Z: X1 = X: T1 = 0
EL = EL0: EM = 0: EN = SQR(1 - EL * EL - EM * EM)
FOR J = 1 TO 3
ZV = Z - T(J)
D = 1 - (S(J) + 1) * EN * EN
E = C(J) * (X * EL + Y * EM - S(J) * ZV * EN) - EN
F = C(J) * (X * X + Y * Y - S(J) * ZV * ZV) - 2 * ZV
SD = F / (SQR(E * E - D * F * C(J)) - E)
X = X + SD * EL
Y = Y + SD * EM
Z = Z + SD * EN - T(J)
T1 = T1 + T(J)
Z2 = Z + T1: X2 = X
LINE (Z1, X1)-(Z2, X2), 9
```

```
Z1 = Z2: X1 = X2
OP = 1 + S(J) * C(J) * Z
SQ = SQR((X * X + Y * Y) * C(J) * C(J) + OP * OP)
COSTHETA = (EN * OP - C(J) * (EL * X + EM * Y)) / SQ
BC = N(J) * N(J) * (1 - COSTHETA * COSTHETA) / NP(J) / NP(J)
KOVERC = NP(J) * SQR(1 - BC) - N(J) * COSTHETA
K = C(J) * KOVERC
EL = (N(J) * EL - K * X / SQ) / NP(J)
EM = (N(J) * EM - K * Y / SQ) / NP(J)
EN = (N(J) * EN + KOVERC * OP / SQ) / NP(J)
NEXT J
EL0 = EL0 + .1
IF EL0 < .9 THEN 10

T1 = 0
FOR Q = 1 TO 3
T1 = T1 + T(Q)
X8 = 0: Z8 = T1
FOR H = 1 TO 30
X = X8 + 3: Z = X * X * C(Q) / (1 + SQR(C(Q) * C(Q) * S(Q) * X * X + 1))
X9 = X: Z9 = Z + T1
LINE (Z8, X8)-(Z9, X9)
X8 = X9: Z8 = Z9
NEXT H
X8 = 0: Z8 = T1
FOR H = 1 TO 30
X = X8 - 3: Z = X * X * C(Q) / (1 + SQR(C(Q) * C(Q) * S(Q) * X * X + 1))
X9 = X: Z9 = Z + T1
LINE (Z8, X8)-(Z9, X9)
X8 = X9: Z8 = Z9
NEXT H
NEXT Q
```

PROBLEM 77

Compare graphically the behavior of a parabolic and a spherical mirror with the same curvature.

PROBLEM 78

A *Ritchey-Chrétien* telescope uses two hyperbolic mirrors arranged as shown in Figure 4.5. The large *primary* mirror on the right has a small hole at the vertex to permit observation or photography. The *secondary* mirror on the left is made as small as possible to avoid reducing the amount of light the telescope can collect. The constants are

Vertex curvatures	–1/8, –1/3, 0
Shape factors	0.07495424, 4.01267321, 0
Mirror spacing	3

Write a program to reproduce this figure.

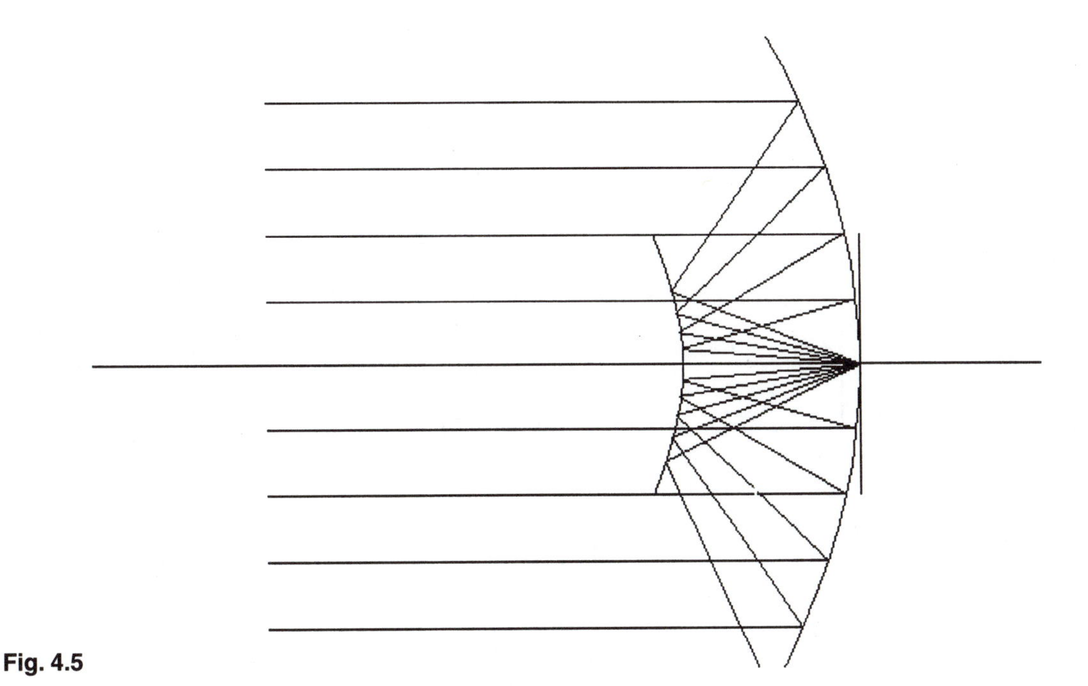

Fig. 4.5

4.5 A LENS WITH AN ELLIPSOIDAL SURFACE

Consider a glass ellipsoid with index n with a point source of light at the focus F_1. A ray F_1P_1 emerges as shown in Figure 4.6, and after refraction it is parallel to the axis. The travel time to an arbitrary image plane at P_2 is

$$t = T_1/v + T_1'/c = (T_1 + T_1'/n)/v$$

where c is the velocity in air, v is the velocity in glass, and $n = c/v$. The quantity t, by Fermat's principle, is a maximum or minimum, but, if the two are identical, then we can regard t as a constant. An ellipse is a conic section for which the sum of the distances from F_1 to P_1 and from P_1 to P_2 are constant, and the eccentricity e of the ellipse—defined as the ratio of the distances P_1F_1 to T_1'—is yet another constant. That is, $T_1 + P_1F_2$ or $T_1 + eT_1'$ is constant, and, comparing this to the expression for t, we see that $e = 1/n$. Hence, a point source at the focus, more remote from the vertex where the rays are leaving and with the right value for the index of the glass, results in a system free of spherical aberration. This arrangement is not of much practical use, but, if we construct a lens whose first surface is spherical with the center at the point source, all rays will strike it normally and be unaffected. Then the second surface can be an ellipsoid, and the lens produces a parallel beam.

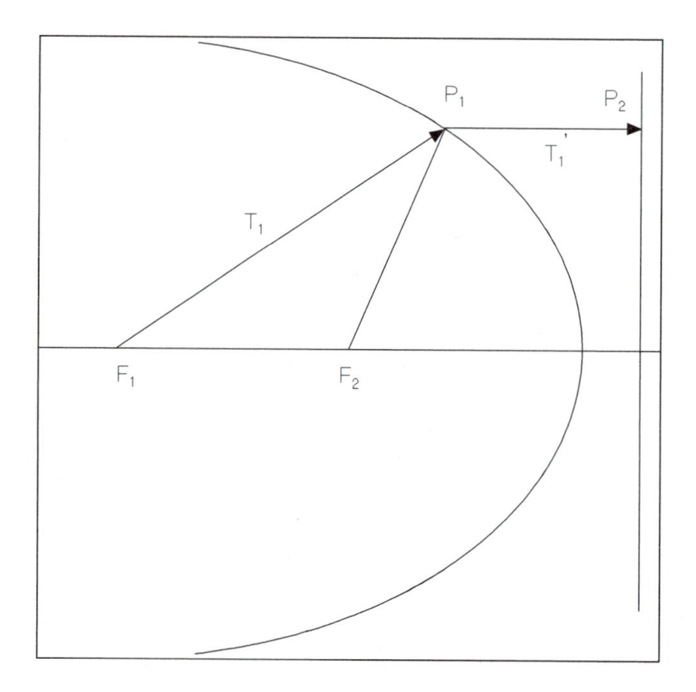

Fig. 4.6

PROBLEM 79

(a) Consider a transparent ellipsoid with vertex V_1, foci F_1 and F_2, and vertex V_2, in that order. Show from (168) and (169) that the distance from F_1 to the more remote vertex V_2 is

$$\overline{F_1 V_2} = \frac{\sqrt{1 + S} + 1}{CS} \tag{179}$$

(b) Figure 4.7 shows a lens and a point source that is the center of the first surface and the focus F_1 for the second surface. The first surface is a sphere of radius 50, and the second surface is an ellipsoid for which the curvature is –0.02 and the shape factor is –0.75. Verify that the eccentricity is 0.5, the index is 2.0, and the lens thickness is 50. Perform the ray tracing.

4.6 The Plano-Hyperbolic Lens

Similar arguments can be applied to a hyperbolic refracting surface. It is shown in analytic geometry books that the hyperbola is the locus of all points for which the difference between the distances to the two foci is constant (rather than the sum, as for the ellipse). From this property, it is found that the index should be equal to the eccentricity instead of to its reciprocal. To construct a lens that acts like the one just described, we place a point source at one focus and have it strike the surface of the hyperbola associated with the other focus, as shown in Figure 4.8. The parallel rays

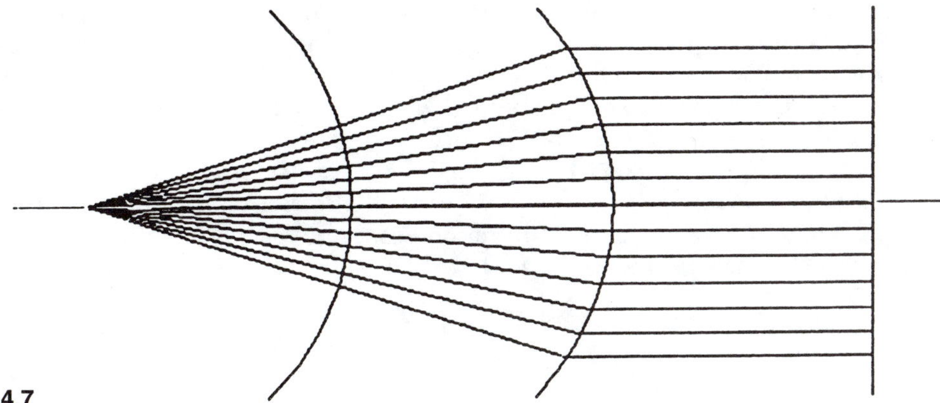

Fig. 4.7

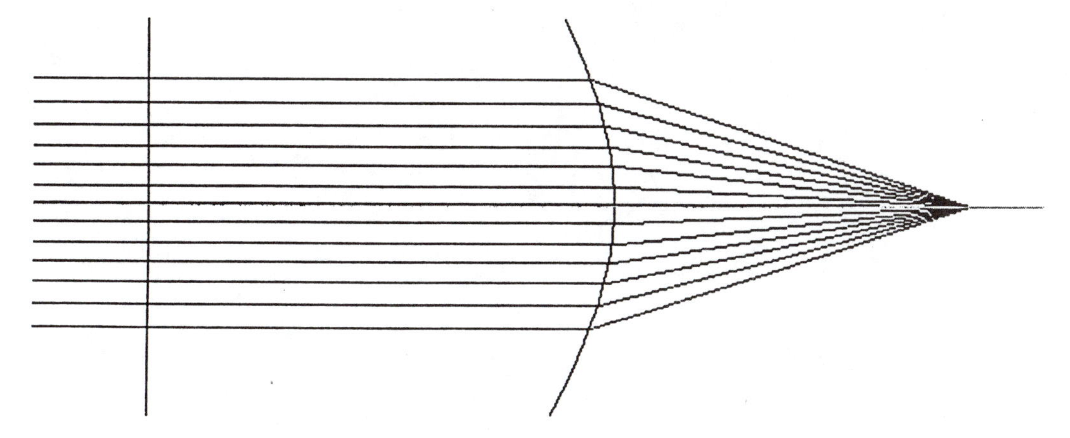

Fig. 4.8

then pass without deviation through a plane second surface, whose position can be in any convenient location.

PROBLEM 80

The lens of Figure 4.8 has a vertex curvature $C = 0.02$, a shape factor $S = 2.0$, and thickness $t_1' = 90.0$. Using equation (179) to find the distance from the point source to the vertex, confirm the ray tracing shown.

4.7 Conic Cylinders

We now consider systems that do not have symmetry about the z-axis. An important example is the parabolic trough used as a collector of solar energy. Parallel rays from the sun are brought to a focus on long, thin heat-exchanger tubes. If the rays are mostly meridional, the ray tracing is a two-dimensional procedure like many of the examples in this chapter. However, when the sun comes in at an angle, it is necessary to trace skew rays. We shall explain how to deal with this problem by introducing the modification of our aspheric skew procedure as given by N. Lessing, *Journal of the Optical Society of America* 52, 472 (1962). Place a cylinder with its axis parallel to OY. Let a ray undergo a displacement $\mathbf{T}_1$ and consider its projection $T_1{}^*$ on the x,z-plane, which is the plane normal to the y-axis. Figure 4.9 shows the original and the projected displacements. The two sets of direction-cosine of the ray must each have the sum of their squares be equal to unity, so that

$$L_1{}^{*2} + N_1{}^{*2} = 1$$

and

$$L_1{}^2 + M_1{}^2 + N_1{}^2 = 1$$

which will be simultaneously satisfied if

$$L_1{}^* = L_1/(1 - M_1{}^2)^{1/2}$$

and

$$N_1{}^* = N_1/(1 - M_1{}^2)^{1/2}$$

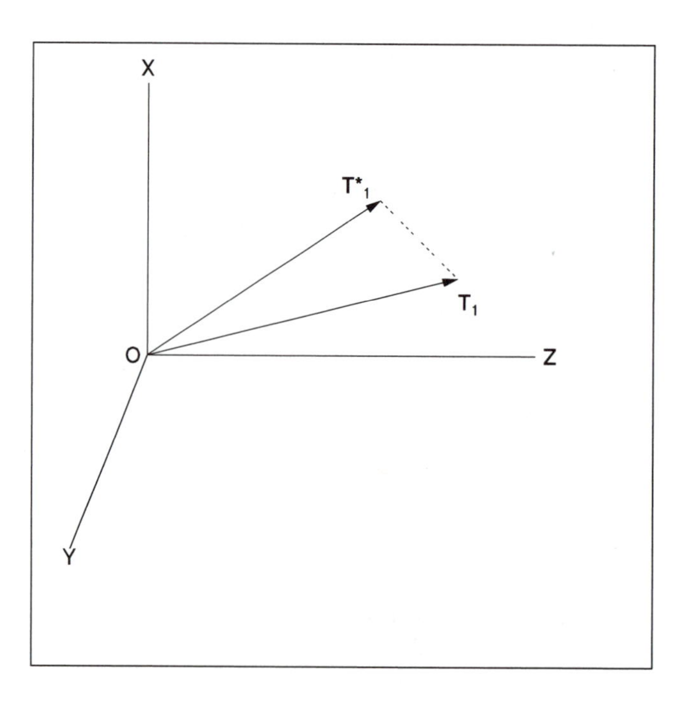

Fig. 4.9

The projection T_1^* of the ray has a length given by the product of T_1 and the cosine of the angle it makes with the x,z-plane. This cosine is the sine of the angle between $\mathbf{T}_1$ and OY, so that

$$T_1^* = T_1(1 - M_1^2)^{1/2}$$

For $y = 0$, equations (172)–(174) are then altered to

$$D^* = 1 - (S_1 + 1)N^2/(1 - M^2)$$

$$E^* = C_1[\{(zN + xL) - (S_1 + 1)zN + S_l v_l N\} - N]/(1 - M^2)^{1/2}$$

$$F^* = C_1[x^2 + z^2 - (S_1 + 1)z^2 + S_1 v_1(2z - v_1) + 2(v_1 - z)$$

Then the length of the ray is

$$T_1 = [F^*/(1 - M^2)^{1/2}]/[-E^* + (E^{*2} - D^*F^*C_1)^{1/2}]$$

and this value of T_1 is used in the translation matrix equations.

For the refraction process, the terms for which $y = 0$ will also drop out, and this has the effect of changing the associated direction-cosine equation to

$$M_1' = n_1 M_1 / n_1'$$

which is simply Snell's law for a plane interface. This is exactly what we would expect for a cylinder parallel to the y-axis.

PROBLEM 81

A double convex cylindrical lens has radii of 50 and –50, a thickness of 15, and an index of 1.5. The first vertex is 200 units from the origin. Trace a ray from the point x = 10, y = 5, and z = 0 with direction-cosines $L = -0.1$ and $M = 0.1$ to its intersection with the paraxial focal plane. In particular, confirm the expression for M_1.

PROBLEM 82

The famous *Hubble space telescope* is a Ritchey-Chrétien with an aperture of 2.4 m. According to an article by Bruce Walker in the *SPIE Newsletter* (November 1990), it has the following parameters.

Shape factors	0.00223, 0.49679
Radii	11.04 m, 1.37 m
Mirror spacing	4.9 m

(Following the custom of astronomers, the article specifies the shape factors as *Schwarzschild* constants *SC*, where $S = -[SC + 1]$.) Verify the ray tracing of Figure 4.10 and locate the paraxial focus.

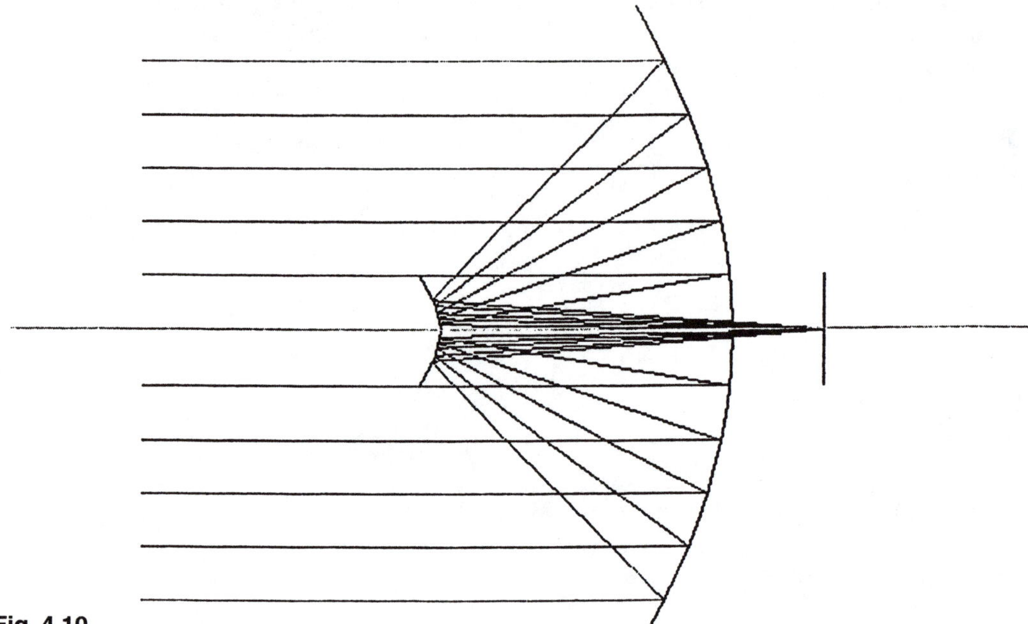

Fig. 4.10

CHAPTER 5

More about Aberrations

5.1 TILTED ELEMENTS

When lenses are improperly mounted, rays that are supposed to be meridional may become skew, and unexpected aberrations are introduced. A component will be tilted or decentered if its axis does not coincide with the symmetry axis of the system. To show what can happen, we shall trace a ray through a tilted element, using an example of R. Kingslake (*Lens Design Fundamentals*, Academic Press, [1978]). A cemented doublet is specified as shown in Table 5.1. A meridional ray 2 units above the axis and parallel to it starts at 2 units from the first vertex and is traced to the paraxial image plane at $t_3' = 11.286$, with the following results.

Table 5.1 Specifications for a Cemented Doublet

c	t'	n'
0.1353271		
	1.05	1.517
–0.1931098		
	0.40	1.649
–0.0616427		

$x' = 0.001366 \qquad z' = 0.0$

Now consider the hypothetical possibility that just surface 3 could pivot about its vertex through an angle of −1° (i.e., clockwise). The point of intersection will be displaced and x_1 and z_1 will be altered. To compute the new coordinates, we need the equation of a circle whose axes have been rotated about the origin. This is done with a rotation matrix, using Figure 5.1. Consider a point P with coordinates z, x. Let a second set of axes Z′OX′ be generated by rotating ZOX through a negative angle ϕ. The figure shows that

$$z' = z \cos\phi - x \sin\phi$$

$$x' = z \sin\phi + x \cos\phi$$

which in matrix form is

$$\begin{pmatrix} z' \\ x' \end{pmatrix} = \begin{pmatrix} \cos\phi & -\sin\phi \\ \sin\phi & \cos\phi \end{pmatrix} \begin{pmatrix} z \\ x \end{pmatrix} \tag{180}$$

By (158), a circle in the Z′OX′ coordinate system, where OZ′ is the horizontal axis, will have the equation

$$x'^2_3 + z'^2_3 - 2z'_3/c_3 = 0$$

To determine the equation of the tilted surface in the ZOX coordinate system, substitute for z'_3 and x'_3 to obtain

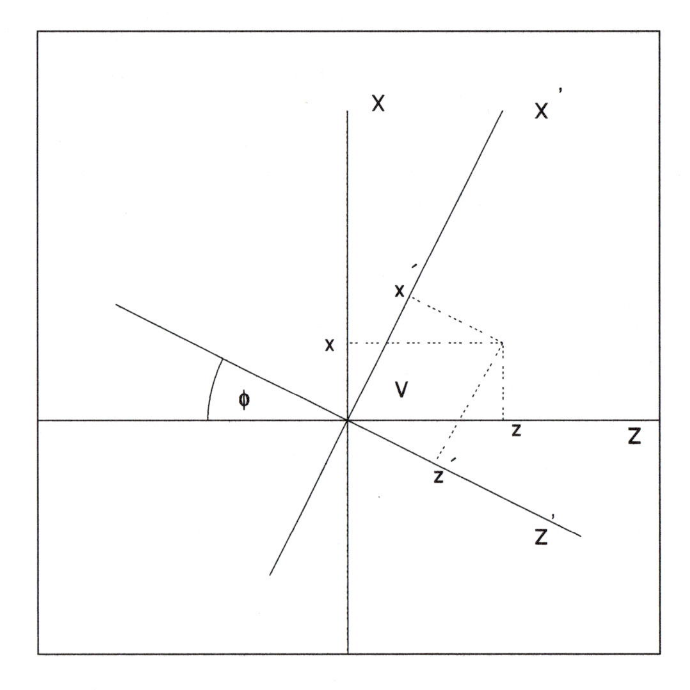

Fig. 5.1

$$x_3^2 + z_3^2 - \frac{2}{c_3}(z_3\cos\phi - x_3\sin\phi) = 0 \tag{181}$$

as the equation of the tilted circle. The final term in (181) changes the quadratic for T_1, as previously given in chapter 2, and the coefficients at surface 3 become

$$E = c_3[x_3L_3 + (z_3 - v_3)N_3] - N_3\cos\phi + L_3\cos\phi$$

$$F = c_3[x_3^2 + (z_3 - v_3)^2] + 2[(v_3 - z_3)\cos\phi - x_3\sin\phi]$$

Note that, when $\phi = 0$, these reduce to the previous expressions for E and F, as they should.

The refraction involves finding the direction-cosines of the normal. In Chapter 4 it was shown that these quantities could be obtained by finding the components of the gradient. Write (181) as

$$f(z_3,x_3) = \frac{c_3}{2}(x_3^2 + z_3^2) - z_3\cos\phi + x_3\sin\phi \tag{182}$$

and differentiate to obtain

$$\frac{\partial f}{\partial x_3} = c_3x_3 + \sin\phi, \quad \frac{\partial f}{\partial z_3} = c_3z_3 - \cos\phi \tag{183}$$

The matrix equations then become

$$L_3' = (n_3/n_3')L_3 - K_3(x_3 + \sin\phi)/c_3n_3'$$

$$N_3' = (n_3/n_3')N_3 - K_3(c_3z_3 - \cos\phi)/c_3n_3'$$

In addition, the components of the normal given above will change the expression for the Snell's law angle to

$$\cos\theta = -L_3(c_3x_3 + \sin\phi) - N_3(c_3z_3 - \cos\phi)$$

and these relations also reduce to the previous expressions for $\phi = 0$. We find that the ray under consideration will strike the paraxial focal plane at a distance $x' = -0.15$, which measures the shift of the ray due to tilting.

PROBLEM 83

Verify the formulas and the ray tracing given above.

PROBLEM 84

Give a double convex lens a 1° tilt. Compare coma before and after tilting.

5.2 TELESCOPE MIRRORS

Spherical mirrors for telescopes are relatively easy to grind. An amateur astronomer will start with a glass disk about 6 in. in diameter and 1 in. thick and produce a spherical surface by using a convex tool and successively finer abrasives. Since a pair of spheres—one concave and the other convex—are the only surfaces that fit together regardless of orientation, the person doing the grinding merely has to walk completely around the blank a large number of times, exerting uniform pressure. But the final product will suffer from spherical aberration; a paraboloidal shape is what is really wanted. Telescope makers have known how to selectively polish spherical mirrors to change them to paraboloids. To see what is done, write (158) as

$$CSz^2 + 2z - Cx^2 = 0$$

Solving for z

$$z = [-1 \pm (1 + C^2Sx^2)^{1/2}]/CS$$

Choosing the positive sign in front of the radical, and multiplying top and bottom by $[1 + (1 + C^2Sx^2)^{1/2}]$ converts this to

$$z = Cx^2/[1 + (1 + C^2Sx^2)^{1/2}]$$

For $C = 0$, the use of the negative sign in front of the radical would lead to division by zero, so that the positive sign is the correct choice. Using the binomial theorem gives

$$(1 + C^2Sx^2)^{1/2} = 1 + {}^1\!/_2C^2Sx^2 + {}^1\!/_2(-{}^1\!/_2)(1/2!)(C^2Sx^2)^2 + \ldots$$

so that

$$z = {}^1\!/_2Cx^2 - (1/8)C^3Sx^4$$

retaining only two terms of the binomial expansion. If $S = 0$, this equation specifies a parabola of the form

$$z = {}^1\!/_2Cx^2$$

as it should. The value $S = -1$ defines a circle, and z becomes

$$z = Cx^2/2 + C^3x^4/8$$

The first term would produce a parabola, and the second term—being of the same sign as the first—moves the edges farther from the vertical axis. Thus, a circle and a parabola with a common curvature are related as shown in Figure 4.3. This figure uses a common value of C for all the conic sections shown, and it demonstrates that they are virtually identical in a region very close to the axis.

PROBLEM 85

Reproduce Figure 4.3 for reasonable values of C and S.

To convert a spherical mirror to parabolic shape, remove material; the farther from the center, the more polishing is necessary. But this can be reduced if the telescope maker is willing to accept a suitable approximation, namely to have the mirror coincide precisely with a parabola only at the center (x = 0) and at the rim ($x = h$). They do this by specifying the parabola as

$$z = C'x^2/2$$

at x = 0, and for $x = h$

$$C'h^2 = Ch^2 + C^3h^4/4$$

or

$$C' = C + C^3h^2/4$$

and the parabola has the form

$$z = (Cx^2/2)[1 + (Ch/2)^2]$$

The parabola and the circle will then be separated by a horizontal distance d given by the difference between this expression and the one given earlier for the circle, so that

$$d = (C^3/8)(x^2h^2 - x^4)$$

Differentiating, the maximum separation occurs at

$$x_{max} = \frac{h}{\sqrt{2}} \tag{184}$$

and this separation has a value

$$d_{max} = C^3h^4/32$$

which is one-fourth as much as originally required. If you are wondering whether an approximate parabola is as good as a perfect one, a mirror whose radius of curvature is 100 in. will lead to a value of d_{max} somewhat less than the wavelength of light. Although the transition from spherical to paraboloidal shape removes the spherical aberration of a telescope mirror, the coma still represents a problem. One fairly effective way of reducing coma was devised by Ross (*Astrophysics Journal* 81, 156 [1935]), using some very complicated theory. Kingslake, in his book on lens design, shows how to determine the specifications of a Ross corrector using a two-step iteration, obtaining the air-spaced doublet specified in Table 5.2.

Unusual features of this system are that the first surface is flat, the two components are very close together, and both are made from the same kind of glass. Kingslake places this corrector at 90 units from the parabolic mirror and close to the image plane (Figure 5.2). The original comatic pattern of the mirror is shown in Figure 5.3, and Figure 5.4 is the pattern after the corrector is incorporated into the telescope. Using the window area on the screen as a measure of the amount of coma, the corrector has reduced the coma noticeably.

Table 5.2 Specifications for an Air-Spaced Doublet

c	t'	n'
0		
	0.3	1.52031
0.1169		
	0.149348	
0.0670		
	0.65	1.52031
−0.0576113		

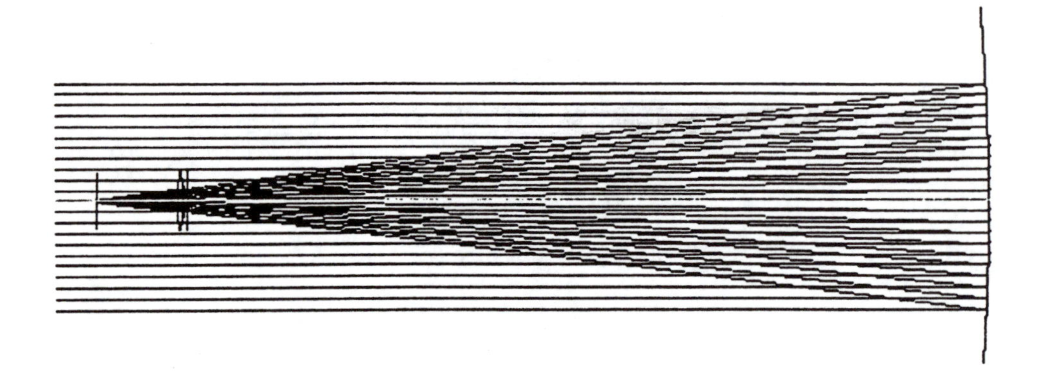

Fig. 5.2

PROBLEM 86

Write a program to alter the position of the Ross corrector from the keyboard, and use it to display the coma pattern for the telescope specified in Table 5.2 and to show that the coma can be further reduced by raising the distance from mirror to corrector by about 0.5 units. *Caution*: The light will be traveling from right to left through the Ross corrector; take this into account for the signs of the surface curvatures.

5.3 THIRD-ORDER ABERRATION THEORY

We can develop a theoretical treatment of the geometrical aberrations if we recognize that light is a wave phenomenon, as we shall consider in more detail in Chapter 7. Let a parallel beam strike a perfect lens and come to a focus at F′. Using this focal point as a center and constructing a sphere S (Figure 5.5), every ray emerging from the lens will strike it perpendicularly. This normal surface is called a *wave front*, and the particular surface that is just emerging from the lens is the *reference sphere*. If the lens

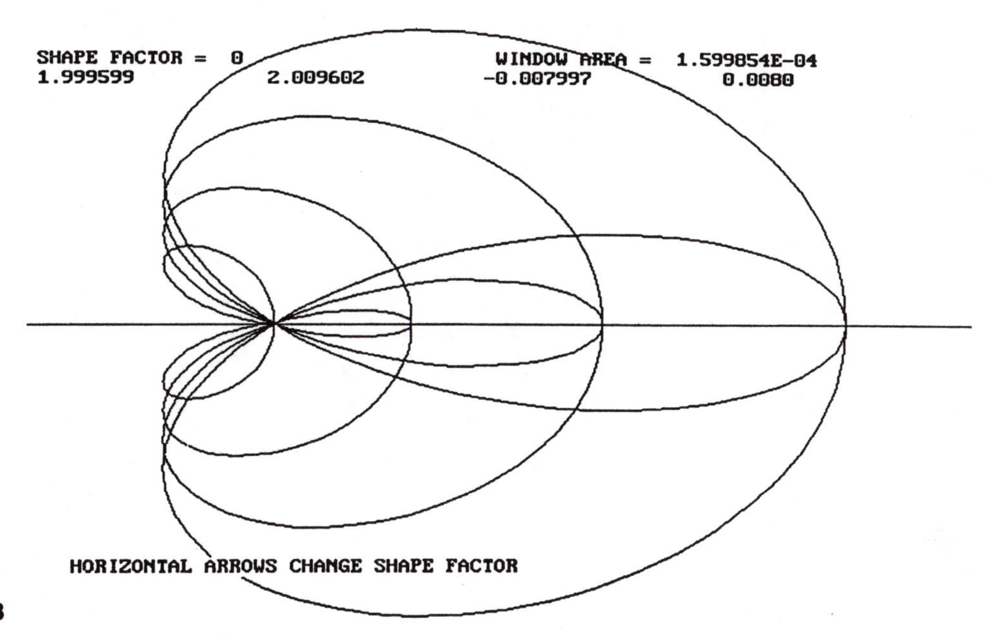
SHAPE FACTOR = 0
WINDOW AREA = 1.599854E-04
1.999599
2.009602
-0.007997
0.0080
HORIZONTAL ARROWS CHANGE SHAPE FACTOR

Fig. 5.3

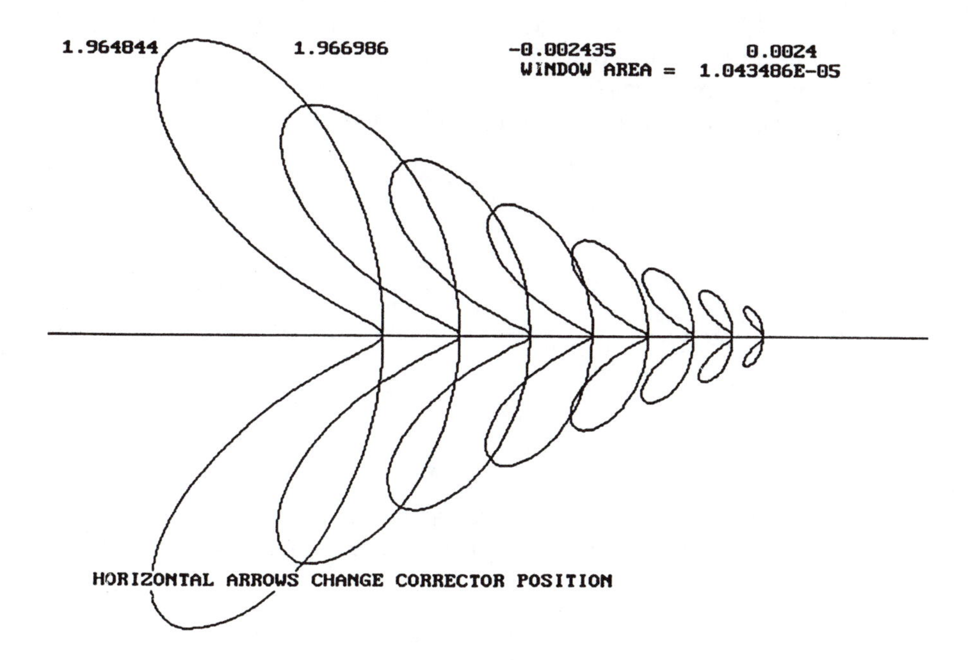
1.964844
1.966986
-0.002435
0.0024
WINDOW AREA = 1.043486E-05
HORIZONTAL ARROWS CHANGE CORRECTOR POSITION

Fig. 5.4

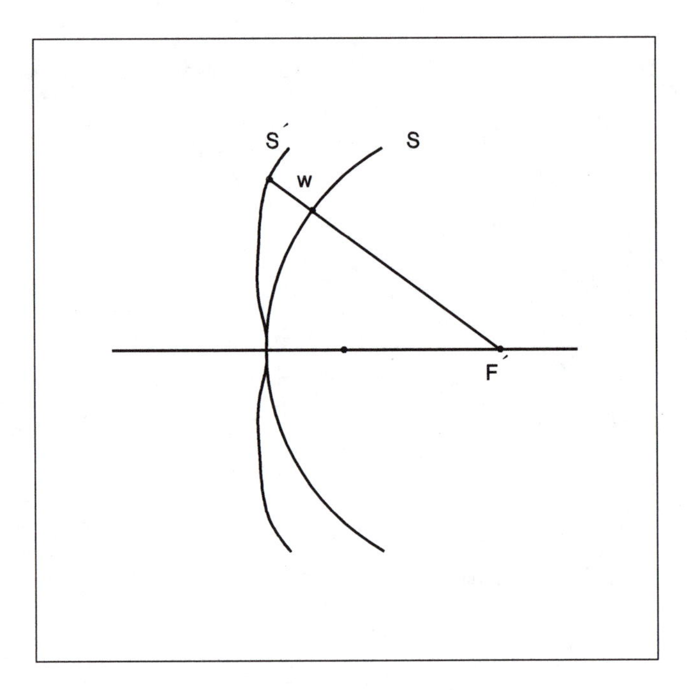

Fig. 5.5

has aberrations, the emerging surface will not be a perfect sphere; instead it will be the distorted surface S′, and the rays perpendicular to it will not meet at a point. The spacing between the reference sphere and the distorted surface is called the *wave aberration w*. We determine w by using Figure 5.6, which shows a reference sphere of radius R and center at the point P′ with coordinates x', y', z'.

For convenience, let S pass through the origin, and choose a point on the distorted sphere S' lying in the x,y-plane and designated $P(x', y')$. The distance w is then

$$w = R - \overline{P'P} \tag{185}$$

or

$$w = \sqrt{x'^2 + y'^2 + z'^2} - \sqrt{(x' - x)^2 + (y' - y)^2 + z^2} \tag{186}$$

which can be written as

$$w = z'\left[\sqrt{1 + \frac{x'^2 + y'^2}{z'^2}} - \sqrt{1 + \frac{(x' - x)^2 + (y' - y)^2}{z'^2}}\right] \tag{187}$$

To put this into a more useful form, we assume that z' is large compared to x' and y', so that the two radicals can be expanded by the binomial theorem, which we use in the form

$$(1 + a)^{1/2} = 1 + a/2 - a^2/8$$

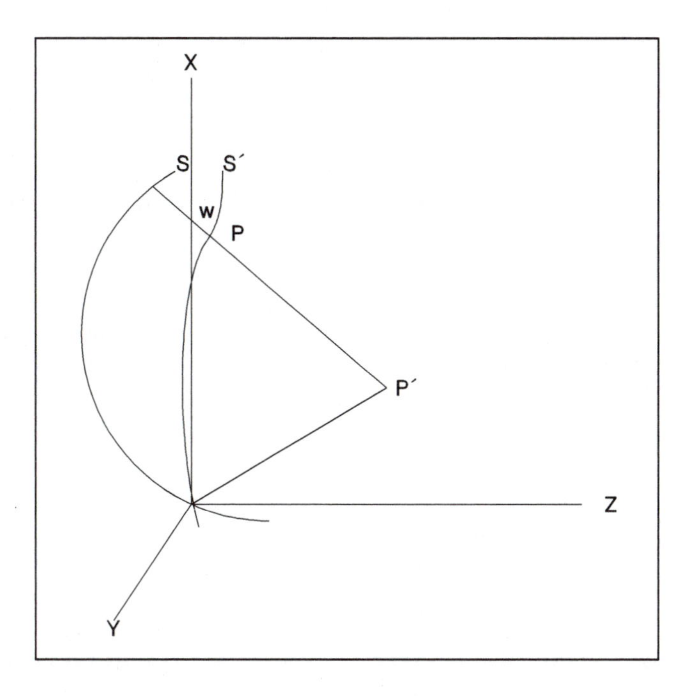

Fig. 5.6

Without loss of generality, we can put P in the x,z-plane, making y′ = 0. Also, let $x' = h$, where h will be the image height as the distorted sphere proceeds toward P′. Combining these changes of notation with the binomial approximation, (187) becomes

$$w = \frac{1}{2z^2}\left[2xh - x^2 - y^2\right] + \frac{1}{8z^4}\left[(x^2 + y^2)^2 - 4hx^3 - 4hxy^2 + 6h^2x^2 + 2h^2y^2 - 4h^3x\right] \quad (188)$$

Now express the points on the emerging wave front in polar coordinates as

$$x = \rho \cos \phi$$

$$y = \rho \sin \phi$$

Making these substitutions and eliminating sin ϕ in favor of cos ϕ (for a reason to be given below), the wave aberration becomes

$$w = \frac{2h\rho \cos\phi - \rho^2}{2z^2} + \frac{\rho^4 + 4h\rho^3\cos\phi - 6h^2\rho^2\cos^2\phi - 2\rho^2h^2(1 - \cos^2\phi) + 4h^3\rho\cos\phi}{8z^4} \quad (189)$$

This equation expresses w as a power series in the variables h, ρ, and ϕ, and it is usually written as

$$w(h,\rho,\phi) = [C_{020}\rho^2 + C_{111}h\rho\cos\phi] + [C_{040}\rho^4 + C_{131}h\rho^3\cos\phi + C_{222}h^2\rho^2\cos^2\phi + C_{220}h^2\rho^2 + C_{311}h^3\rho\cos\phi] \quad (190)$$

where the subscripts on the constants match the exponents on h, ρ, and $\cos\phi$, respectively. Having gone through the horrendous algebra involved in deriving this expression, we now show that we could have guessed its form in an intuitive way. Letting $y_0 = 0$ means that we are assuming that there is symmetry about the y,z-plane. We would expect that the wave aberration for a given ray will depend on where it leaves the lens, as specified by its polar coordinates ρ and ϕ, and it will also depend on the image position h. The angular dependence must show symmetry, so that it has to be expressed in terms of $\cos\phi$, an even function. If $h = 0$, then the image is on the axis, and the system is completely symmetrical about OZ. Therefore, terms not containing h cannot contain $\cos\phi$, and they can only involve even powers of ρ. If a term contains h or h^3 it must also contain ρ or ρ^3. Using these and similar arguments, we find that (189) involves all allowed combinations of the three variables.

Next we consider the nature of the image. Let us write the equation of the distorted sphere as

$$x^2 + y^2 + z^2 = (R + w)^2$$

or

$$x^2 + y^2 + z^2 - R^2 - 2Rw = 0$$

where we have neglected the term w^2. Then differentiating with respect to x

$$2x - 2R\frac{\partial w}{\partial x} = 0 \tag{191}$$

or

$$\frac{\partial w}{\partial x} = \frac{x}{R} \tag{192}$$

and similarly

$$\frac{\partial w}{\partial y} = \frac{y}{R} \tag{193}$$

When the distorted sphere reaches the image position, the coordinates in this plane are designated as x' and y'. Then (192) and (193) become

$$x' = R\frac{\partial w}{\partial x}, \quad y' = R\frac{\partial w}{\partial y} \tag{194}$$

It is useful to have polar forms of these equations, obtained by differentiating the definition $x = \rho\cos\phi$ to obtain

$$1 = \frac{\partial \rho}{\partial x}\cos\phi - \rho\sin\phi\frac{\partial \phi}{\partial x} \tag{195}$$

which is an identity if

$$\cos\phi = \frac{\partial\rho}{\partial x}, \quad \sin\phi = -\rho\frac{\partial\phi}{\partial x} \tag{196}$$

as may be verified by substitution. The rules for partial differentiation show that

$$\frac{\partial w}{\partial x} = \frac{\partial w}{\partial\rho}\frac{\partial\rho}{\partial x} + \frac{\partial w}{\partial\phi}\frac{\partial\phi}{\partial x} \tag{197}$$

Using (196) and (197), (194) becomes

$$x' = R\left[\cos\phi\frac{\partial w}{\partial\rho} - \frac{\sin\phi}{\rho}\frac{\partial w}{\partial\phi}\right] \tag{198}$$

and in the same way

$$y' = R\left[\sin\phi\frac{\partial w}{\partial\rho} + \frac{\cos\phi}{\rho}\frac{\partial w}{\partial\phi}\right] \tag{199}$$

At the image plane

$$\rho' = \sqrt{x'^2 + y'^2} = R\sqrt{\left(\frac{\partial w}{\partial\rho}\right)^2 + \frac{1}{\rho^2}\left(\frac{\partial w}{\partial\phi}\right)^2} \tag{200}$$

If w is symmetric about the axis, this simplifies to

$$\rho' = R\frac{dw}{d\rho} \tag{201}$$

We now examine (190) term by term. The first term involves only ρ, so that (200) gives

$$\rho' = 2C_{020}R\rho \tag{202}$$

Thus each ring in the emerging wave produces a ring of radius $2C_{020}R\rho$ at the paraxial plane; the image is not the point that it should be and we have a *longitudinal focusing error.* The other quantity inside the brackets can be written as $C_{111}hx$, so that by (194)

$$x' = RC_{111}h \tag{203}$$

and this is a *transverse focusing error.* These two focusing errors are combined in the first set of brackets on the right-hand side of (190); they are not true aberrations and can be reduced by refocusing. The other pair of brackets encloses five terms; we will soon show that they correspond to the geometric aberrations previously studied. When $\sin\theta$ was approximated by θ, the first term in the Taylor series, no aberrations were present. Including the next term $\theta^3/3!$ should cause aberrations to appear. Rather than work with this complicated algebra, we described the aberrations through the use of exact ray tracing. Here we use a binomial approximation with the higher-order terms dropped, and we conclude that we have obtained the same result that using the Taylor series would give. Thus, we label the defects previously introduced as *third-order aberrations.*

In the second pair of brackets on the right of (190), the only term that does not vanish when h = 0, and which therefore must correspond to an image on the axis, is the term $C_{040}\rho^4$. This means that the coefficient C_{040} is a measure of spherical aberration. Let's combine this with the longitudinal focusing error and write w as

$$w = C_{040}\rho^4 + C_{020}\rho^2 \tag{204}$$

For a double convex lens, spherical aberration moves the marginal focal point closer to the lens, so that making things better for this focal point makes them worse at F′. We can thus say that spherical aberration and the longitudinal focusing error oppose one another. We therefore expect that it would be possible to achieve the best compromise at some plane lying between the paraxial and marginal focal planes; this is what we know as the circle of least confusion.

The wave aberration for the next third-order term is

$$w = C_{131}h\rho^3\cos\phi$$

Substituting it into (198) and (199), we obtain

$$x' = RC_{131}h\rho^2(3\cos^2\phi + \sin^2\phi) = RC_{131}h\rho^2(2 + \cos 2\phi) \tag{205}$$

and

$$y' = RC_{131}h\rho^2\sin 2\phi \tag{206}$$

This pair of equations defines a family of circles: each ring of radius ρ in the emerging wave produces one in the image plane whose center is on the x-axis. As ρ increases, the size of the corresponding image circle increases and the center moves farther away from the z-axis because of the first term on the right-hand side of this equation. The fact that the angle ϕ appears as 2ϕ means that a single ring in the emerging wave generates a double ring in the image. All of these facts mean that this term provides a definition of ideal third-order coma, obtainable when it is the only aberration. If other aberrations are present, coma looks like the cardiod patterns previously found numerically. The next aberration to consider is a little more complicated. Let's combine the terms involving ρ^2 and use rectangular coordinates to obtain

$$w = C_{222}h^2x^2 + C_{220}h^2(x^2 + y^2) \tag{207}$$

Then, by (194)

$$x' = Ax = A\rho\cos\phi, \quad y' = By = B\rho\sin\phi \tag{208}$$

where A and B lump all the constants. This pair of equations define an ellipse, which we recognize as astigmatism. In addition, the term $C_{220}h^2\rho^2$ makes a separate contribution to the group of aberrations—its dependence on both the square of the image height and the size of the emerging wave means that the paraxial focus is shifted both along the axis and normal to it. The image is therefore a series of rings that get successively wider and farther from F′—this is what we have previously called curvature of field.

PROBLEM 87

Show that the last term in the expression for w defines distortion.

We have thus shown the precise nature of each of the third-order aberrations. Note especially that, if we had carried the binomial approximation one step farther, we would have obtained a set of 12 terms that characterize the *fifth-order aberrations*. Although professional lens designers worry about these (and even higher orders), our more elementary approach covers the subject well enough.

CHAPTER 6

Chromatic Aberrations

6.1 The Nature of Chromatic Aberrations

An aspect of lens system design we have not discussed yet is the fact that white light has a range of colors or frequencies and the index of refraction depends on the color. This is demonstrated by the effect of a prism on sunlight. Some typical values of the index n for two types of glass are shown in Table 6.1. The first column in this table identifies the wavelength in terms of its position in the *Fraunhofer spectrum* of the sun. Fraunhofer arbitrarily labeled a number of bright lines in the solar spectrum as A through H; the ones shown correspond to hydrogen (C), sodium (D), and calcium (F). It is possible to express the dependence of the index of refraction on wavelength for a given kind of glass by defining the *dispersive power* or *Abbe number* as

Table 6.1 Typical Values of Index n for Two Types of Glass

Fraunhofer Line	Color	Wavelength (nm)	Index (Crown)	Index (Flint)
C	Red	656.3	1.514	1.694
D	Yellow	589.6	1.517	1.701
F	Blue	486.1	1.522	1.717

$$V = \frac{n_B - n_R}{n_Y - 1} \tag{209}$$

where n_B, n_Y, and n_R are the indices for blue, yellow, and red light, respectively. Listings of lens constants usually give the index for yellow light and the quantity $1/V$; for example, the crown glass column in Table 6.1 gives $1/V = 64.02$. If we are given this number and the index for yellow light, we can estimate the indices for red and blue light by assuming that the yellow index lies halfway between. A calculation based on this approximation makes the red index equal to 1.50885, and the blue index is 1.52447. This is accurate enough when ray tracing for these three different colors to realize that the properties of the lens are wavelength dependent; it will show *chromatic aberration*. Table 6.1 lists two kinds of glass, with *crown* being used, for example, in bottles, windows, and eyeglasses. *Flint* glass has the higher index and is usually harder. The distinction is that crown glasses are those for which the index n_Y is below 1.60 and $1/V$ is greater than 55 or the index is above 1.60 and $1/V$ is greater than 50; all other glasses are flint.

A simple physical interpretation of V is obtained by ray tracing. Figure 6.1 shows what happens at the two ends of the spectrum for a double convex lens. Let the distance from the axis to the edge of the lens (the physical radius) be R. Because of the spread in colors, there is a spread $\Delta f'$ in the focal point and a finite size x of the image. By similar triangles

$$\frac{R}{f'} = \frac{x}{\Delta f'/2} \tag{210}$$

or

$$x = \frac{R\,\Delta f'}{2\,f'} \tag{211}$$

Now write (209) as

$$V = \frac{dn}{n - 1} \tag{212}$$

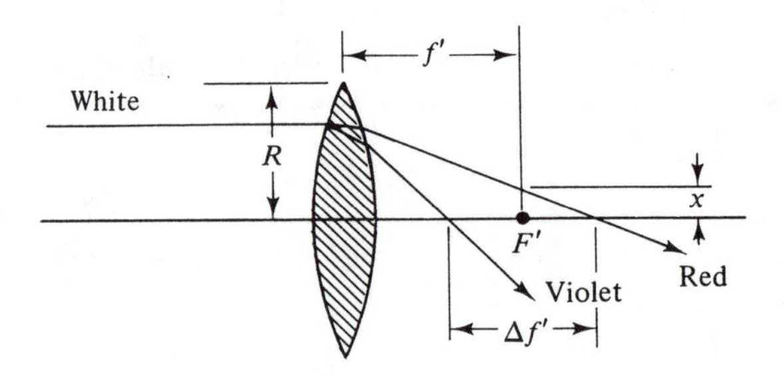

Fig. 6.1

where expressing dn as $n_B - n_R$ is an approximation, which we shall see shortly is not really correct, and where $n = n_Y$ is a sort of an average value of the index of the lens. For a thin lens, we know that the Gaussian constant a is

$$a = \frac{n-1}{r_1} + \frac{1-n}{r_2} \tag{213}$$

When the wavelength changes

$$da = \frac{dn}{(n-1)f'} = d\left(\frac{1}{f'}\right) = \frac{-df'}{f'^2} \tag{214}$$

Combining this with (211) gives

$$x = -\left(\frac{R}{2}\right)V \tag{215}$$

This means that the dispersive power V is a measure of the chromatic aberration, which is the spread of the image due to the fact that different colors focus at different points. Lenses or lens systems that are corrected for this defect are said to be *achromatic*. This is accomplished by making the focal lengths for red light (656 nm) and blue light (486 nm) the same. There will of course still be chromatic errors at intermediate wavelengths, being greatest for yellow light halfway between. For special applications, such as color reproductions in the printing industry, lenses are designed with perfect focusing of yellow as well; these are *apochromatic* lenses.

PROBLEM 88

(a) Show that the chromatic aberration of a thick lens can be minimized by giving the index the value

$$n_1'^2 = \frac{t_1'}{r_2 - r_1 + t_1'} \tag{216}$$

(b) Can such a lens be made?

Two thin lenses can form an achromatic pair if their spacing is properly chosen. An example is the *Huygens eyepiece*, consisting of two plano-convex lenses with the curved surfaces facing toward the objective lens of a telescope. It can be shown, by extending the arguments just given, that the separation of the two lenses should be equal to the average value (that is, half the sum) of their focal lengths. However, it has been pointed out by E. J. Irons (*American Journal of Physics* 31, 940, 1967) that replacing $n_B - n_R$ with an infinitesimal quantity dn is poor mathematics; we shall give a derivation that differs from that found in most optics texts. The formula for the focal length of a thin lens at a specific color, say yellow, is

$$\frac{1}{f_Y} = (n_Y - 1)\left(\frac{1}{r_1} - \frac{1}{r_2}\right) \tag{217}$$

so that at the ends of the spectrum

$$\frac{1}{f_B} - \frac{1}{f_R} = (n_B - n_R)\left(\frac{1}{r_1} - \frac{1}{r_2}\right) = \frac{n_B - n_R}{f_Y(n_Y - 1)} = \frac{V}{f_Y} \tag{218}$$

For a pair of thin lenses spaced a distance t, using (71)

$$\frac{1}{f_B} - \frac{1}{f_R} = \frac{V_1}{f_{1Y}} + \frac{V_2}{f_{2Y}} - t\left(\frac{1}{f_{1B}f_{2B}} - \frac{1}{f_{1R}f_{2R}}\right) \tag{219}$$

If the pair of lenses forms a cemented doublet ($t = 0$) that is achromatic, then the red and blue focal lengths are the same and we obtain the condition

$$\frac{V_1}{f_{1Y}} + \frac{V_2}{f_{2Y}} = 0 \tag{220}$$

This requirement can be satisfied if the two focal lengths are opposite in sign. That is, the cemented doublet can be made from a converging and a diverging lens of the same index. For a spaced doublet, impose the condition that

$$\frac{1}{f_{1Y}} = \frac{1}{2}\left(\frac{1}{f_{1B}} + \frac{1}{f_{1R}}\right) \tag{221}$$

Combining this with (217) as applied to lens 1 gives

$$n_{1Y} = \frac{n_{1B} + n_{1R}}{2} \tag{222}$$

with a similar result for lens 2. These stipulations are approximately true for the glasses in Table 6.1.

Lens designers refer to the image defect associated with red and blue light as *primary chromatic aberration*. When this has been corrected, what is left is *secondary chromatic aberration*. An apochromatic system will reduce this latter type to a very small value.

PROBLEM 89

(a) Let the two lenses in the air-spaced achromatic doublet be identical. Show from (220) and (222) that the spacing is equal to their common focal length.

(b) In M. Laikin, *Lens Design*, Marcel Dekker (1991), the following specifications are given for an apochromatic and virtually aberration-free microscope objective. (*Note:* This book gives the design specifications and resolution curves for an enormous number and variety of optical systems.)

Radius	*Thickness*	*Blue Index*	*Yellow Index*	*Red Index*
0.37308				
	0.0258	1.65992	1.65391	1.64816
0.17458				
	0.1361	1.43728	1.43494	1.43268
–0.16802				
	0.0258	1.65992	1.65391	1.64816
–0.62359				
	0.0099	air		
0.41732				
	0.0606	1.55885	1.55440	1.55008
–0.68980				
	0.0134	air		
0.16865				
	0.1071	1.43728	1.43494	1.43268
–0.17242				
	0.0220	1.63932	1.63004	1.62154
–0.39328				
	0.0099	air		
0.09843				
	0.0773	1.59635	1.59142	1.58666
0				
	0.0083	air		
0				
	0.0071	1.52910	1.52458	1.52024
0				

This objective consists of a cemented triplet, a single positive lens, a cemented doublet, and a nearly hemispherical element with a flat second surface. Then there is a small air space separating the lens system from the cover glass. One of its features is that this cover glass is included in the optical design. Verify the ray tracing diagram of Figure 6.2.

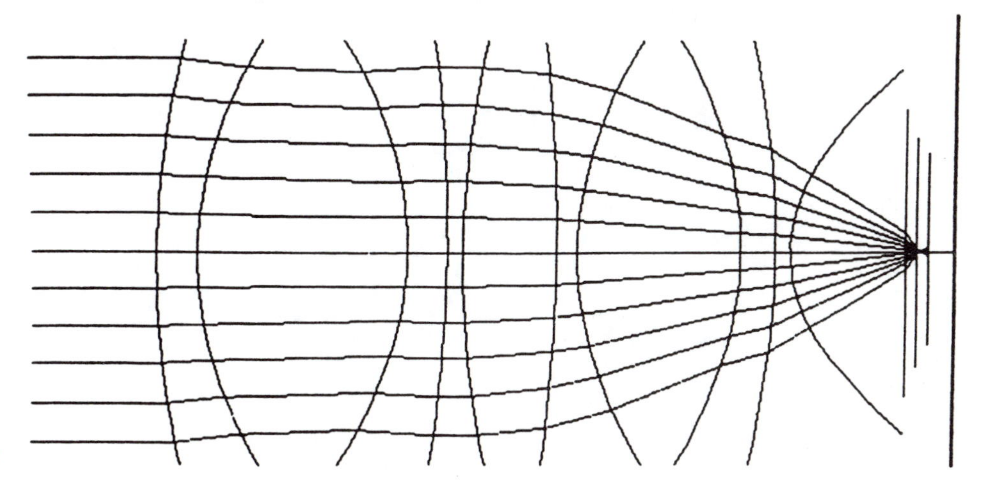

Fig. 6.2

(c) Use the combination of lens and cover glass as a beam spreader. Place a source of laser light at the second surface of the cover glass and have the light go through the objective from left to right in Figure 6.2, emerging as a parallel beam. It is easier to trace parallel rays striking the first element of the objective, and this is how Figure 6.2, using yellow light, was produced. It is seen that the focal point is inside the cover glass. Show how to change the position and thickness of the cover glass to achieve the situation of Figure 6.3.

(d) Then confirm the apochromatic property of the system by tracing red light through the system.

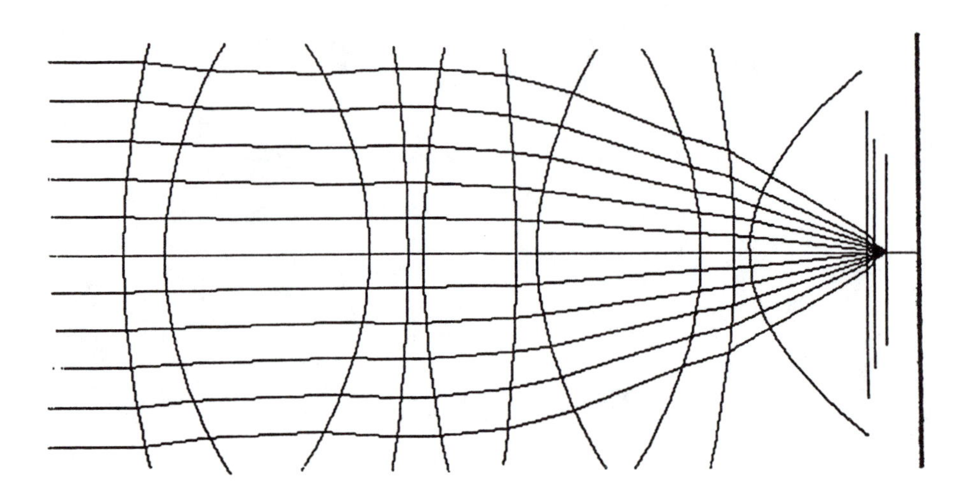

Fig. 6.3

CHAPTER 7

Wave Optics

7.1 Waves and the Classical Wave Equation

Light is the visible part of the *electromagnetic spectrum*, corresponding to wavelengths in the approximate range of 400 to 700 nm. Table 7.1 shows the complete spectrum and the position of visible light in it. The behavior of waves is specified by a differential equation which we shall explain in a very elementary way. Consider a long piece of wire bent into the shape of a cosine curve (Figure 7.1a). It has the equation

$$x = a\cos kz \tag{223}$$

where x is the *amplitude* of the wave with a maximum value of a and k is a quantity to be determined. The distance between two peaks on the curve is the *wavelength* λ, so that when $z = 0$ or λ, then $x = a$. Therefore

$$k\lambda = 2\pi$$

or

$$k = \frac{2\pi}{\lambda} \tag{224}$$

and k is called the *propagation constant* or *wave number*; it is a measure of the number of waves per unit length.

If the rigid curve is made to move to the right with velocity v (Figure 7.1b), it will travel a distance vt in time t, and equation (223) becomes

$$x = a\cos\left[\frac{2\pi}{\lambda}(z - vt)\right] \tag{225}$$

Table 7.1 The Electromagnetic and Acoustic Spectrum

Name	Typical Frequency	Typical Wavelength	Source	Application
DC	0		Battery	Power supplies
Ultralow frequencies	1 Hz	3×10^8 m	Oscillators	Submarine radio
Power line	50 or 60 Hz	6×10^6 m	Hydro	Motors, etc.
Audio	0–30 kHz	30 km	Oscillators	HiFi
Ultrasound	30–400 kHz		Oscillators	Motion detectors
RF	0.5–1.5 MHz	300 m	Oscillators	AM radio
UHF	2–1,000 MHz	1 m	Oscillators	FM radio, TV
Microwave	1–100 GHz	1 cm	TWTs	Satellite signals
IR	10^{11}–4×10^{14} Hz	3,000–0.7 μm	Lasers	Detection
Visible	10^{15} Hz	0.4–0.7 μm	Sunlight	Optics
UV	10^{16} Hz	0.3 μm	Lamps	Germicide
X ray	10^{17} Hz	30 nm	Tubes	Medicine
γ ray	$>10^{19}$ Hz	<0.03 nm		

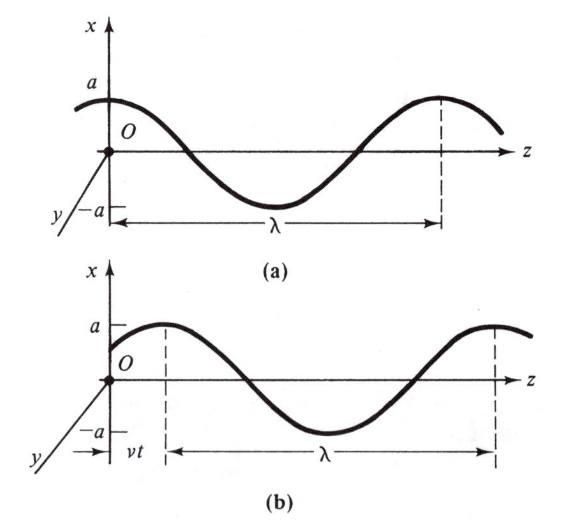

Fig. 7.1

PROBLEM 90

(a) Verify that (225) describes a wave moving to the right.

(b) Show that (225) satisfies the differential equation

$$v^2 \frac{\partial^2 x}{\partial z^2} = \frac{\partial^2 x}{\partial t^2} \tag{226}$$

(c) Show that (225) may be written as

$$x = a \cos(kz - \omega t)$$

where

$$\omega = vk \tag{227}$$

is the *angular frequency* of the wave.

(d) Show that equation (226) is also satisfied by the solution

$$x = a e^{\pm i(kz - \omega t)} \tag{228}$$

7.2 HUYGENS' PRINCIPLE

There is a simple but powerful way of describing waves. Huygens proposed that all points on a wave front are sources of secondary waves. Let's apply Huygens' principle to the wave radiated by the point source P in Figure 7.2. The wave front is a sphere

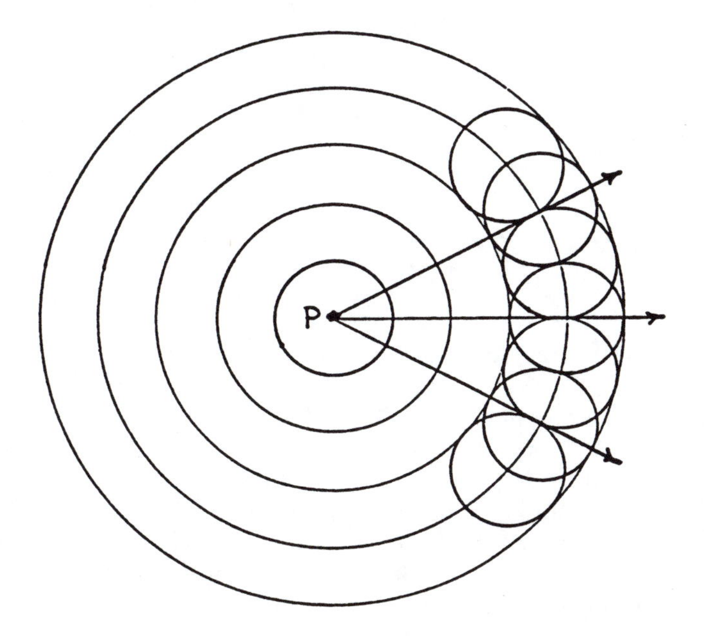

Fig. 7.2

that expands with time, so that it is indicated as a series of concentric circles. Using one of these positions as a source, the smaller spheres shown in the drawing are generated and the tangent surface produces the same kind of wave front as the original. Huygens was able to deduce Snell's law using this approach. Figure 7.3 shows a glass-air interface and a refracted plane wave. Let the wave from point A on the wave front before refraction travel to B in time t. Let a wave from C on the refracted ray travel to D in the same time. Then

$$t = AB/v_1 = CD/v_2$$

so that

$$\sin\theta_1/\sin\theta_2 = v_1/v_2$$

agreeing with the previous form of Snell's law.

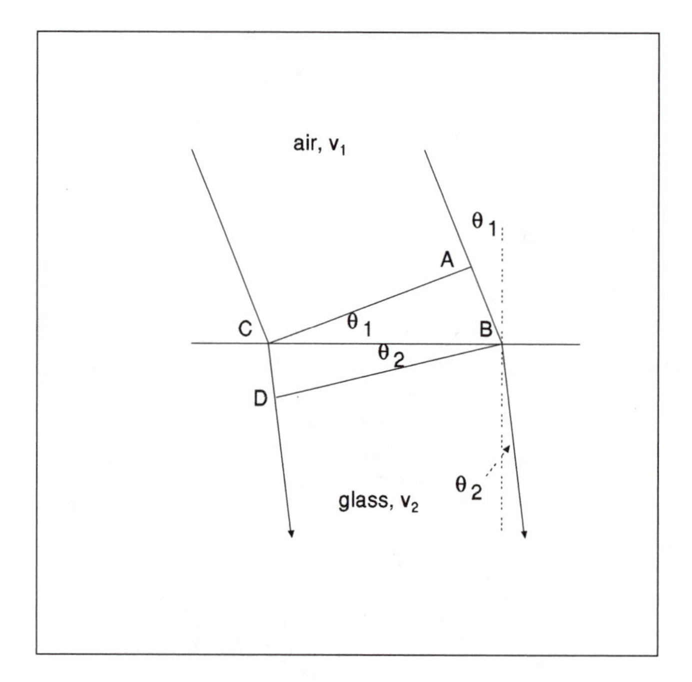

Fig. 7.3

7.3 Fraunhofer Diffraction

Place a metal plate with an opening in the path of a light wave (Figure 7.4). The aperture will act as a Huygens' source of secondary waves that are received on a screen, which should have an illuminated area of the same shape as the opening, while the rest of it is dark. For a small opening, it is found that there are bright regions in places that violate the predictions of geometrical optics. This phenomenon is known as *Fraunhofer diffraction* and its explanation is very complicated. We shall give a simple theory as an application of Huygens' principle. Let us find the amplitude $A(x',y')$ of the

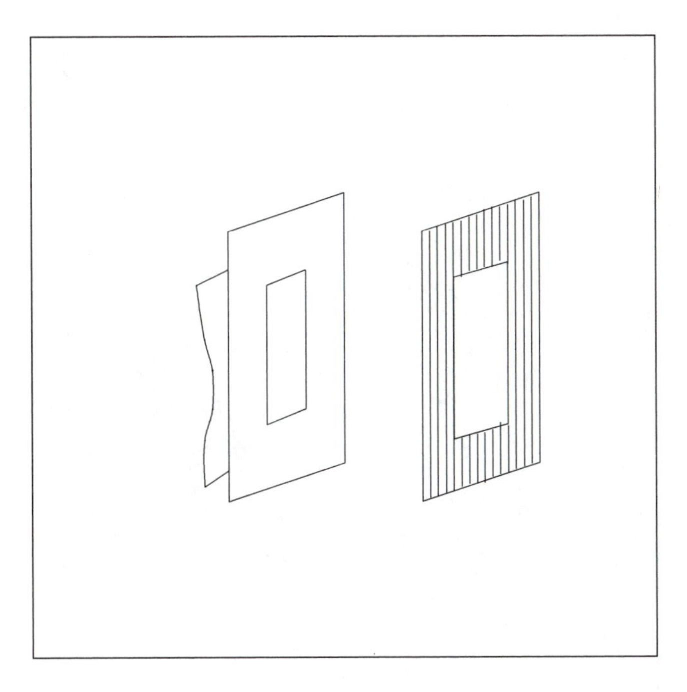

Fig. 7.4

wave received at point P′ on the screen. The light received at P′ can be regarded as the sum of the contributions from each point in the opening. Using equation (228) for the amplitude of the wave, this sum is then expressed as the integral

$$A(x',y') = C\int A(x,y)e^{ikr}\,dS \tag{229}$$

where $A(x,y)$ is the variable amplitude across the aperture and the constant C includes all terms that can be taken out of the integral; it also serves to make the units match on each side of the equation. This result can be regarded as the mathematical expression of Huygens' principle; we shall apply it to rectangular and circular apertures. When this integral is evaluated over an aperture that passes only a portion of an incident wave and blocks the rest, a distant screen receives this radiation with a non-uniform distribution, which we can determine quantitatively.

7.4 DIFFRACTION BY A SLIT

Consider a slit of length l and width w (Figure 7.5), for which w is much greater than l, so that the short edges make no contribution. Two parallel rays leaving the slit at an angle θ are shown in the figure—one from the center of the opening and the other from a point whose coordinates are $(x, 0, 0)$. At a distance r from the slit, we write

$$r = r_0 + x\sin\theta \tag{230}$$

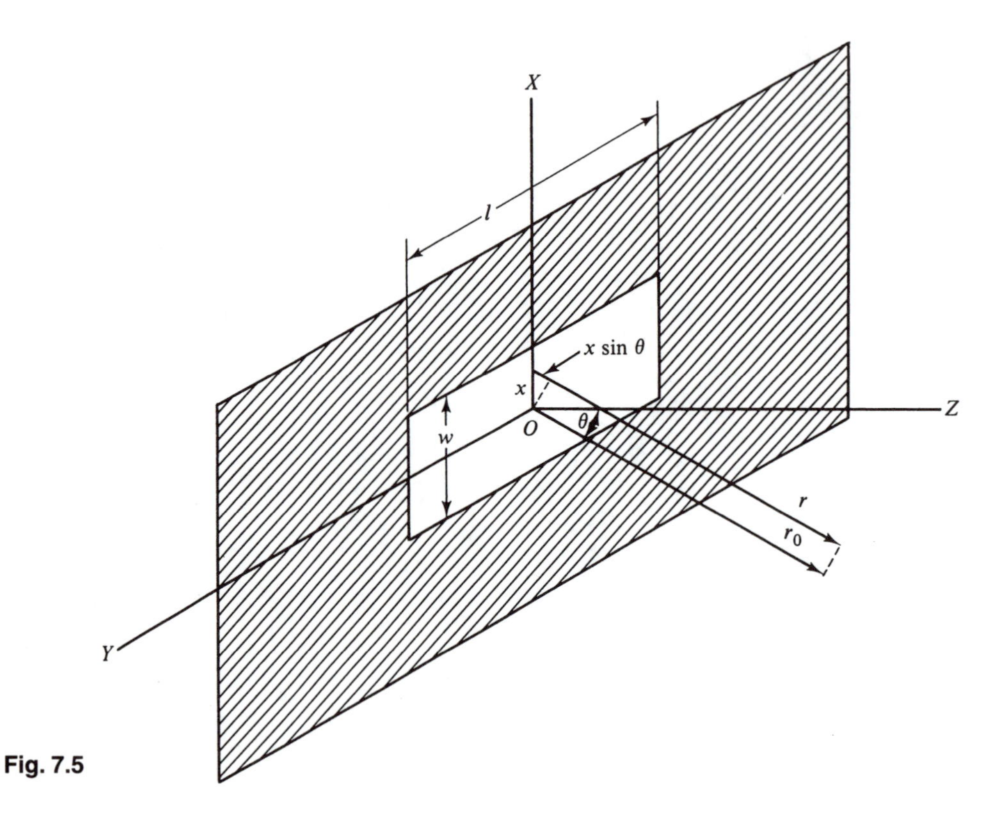

Fig. 7.5

where r_0 and r are indicated by the figure. The element of area in the slit is

$$dS = l\,dx \tag{231}$$

Across small apertures, the term $A(x,y)$ is almost constant. Absorbing this into C, as well as the other constants from inside the integral, (229) simplifies to

$$A = Cle^{ikr_0}\int_{-w/2}^{w/2} e^{ikx\sin\theta}dx \tag{232}$$

Performing the integration gives

$$A = C\left(\frac{\sin\alpha}{\alpha}\right) \tag{233}$$

where

$$\alpha = \tfrac{1}{2}kw\sin\theta \tag{234}$$

This result gives the amplitude as received at the screen, but what the eye detects is the *intensity* I, a measure of the energy in the wave; A measures its strength. For a complex quantity, we obtain a quantity proportional to the intensity by multiplying A by its complex conjugate. Let

$$I_0 = C\,C^* \tag{235}$$

since C is complex, obtaining

$$I = I_0\left(\frac{\sin^2\alpha}{\alpha^2}\right) \tag{236}$$

PROBLEM 91

Confirm (233) and (236).

The curve of I/I_0 vs. α is shown in Figure 7.6. The principal maximum occurs when $I = I_0$, which corresponds to α or θ equal to zero. There are secondary maxima that can be located by differentiating (236), from which we find that

$$\alpha = \tan\alpha \tag{237}$$

This transcendental equation can be solved numerically or graphically to obtain the roots

$$\alpha = 0,\ 1.43\,\pi,\ 2.46\,\pi,\ldots\ldots \tag{238}$$

The minima correspond to $\sin\alpha = 0$, from which

$$\alpha = m\pi,\quad m = \pm1,\pm2,\ldots. \tag{239}$$

where $\alpha = 0$ is not allowed, or

$$\sin\theta = \frac{m\lambda}{w} \tag{240}$$

For small angles, this becomes

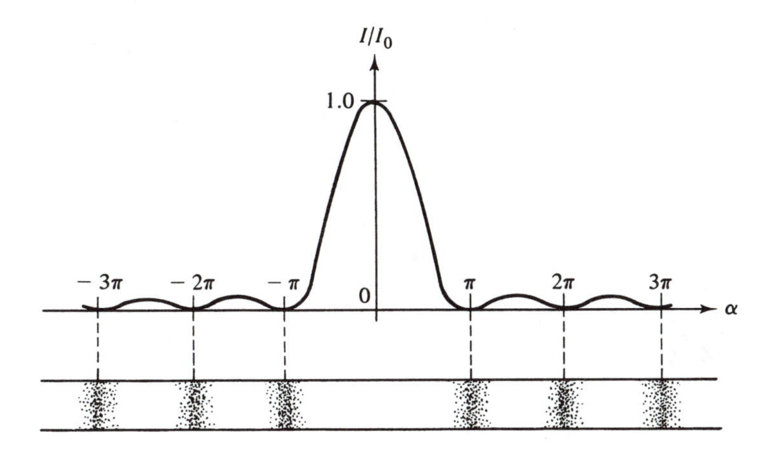

Fig. 7.6

$$\theta = \frac{m\lambda}{w} \tag{241}$$

so that the bright central band has an angular width of $2\lambda/w$. The *diffraction pattern* consists of this intense band, flanked by weaker bands symmetrically arranged, as shown in the lower part of Figure 7.6. If the slit is made narrower, the central band becomes wider and less intense. Increasing w, however, means that both pairs of edges produce a diffraction pattern, and we obtain a two-dimensional rectangular arrangement. It should be noted that the use of Huygens' principle to determine the nature of a diffraction pattern involves approximations that are not easy to justify. For example, the exponential term in the integral involves waves moving in both directions; we have oversimplified the physical situation by considering travel only to the right.

7.5 Diffraction by a Circular Aperture

The geometry for finding the diffraction pattern of a circular opening is shown in Figure 7.7. The element of area is taken as a strip of width dx and length $2\sqrt{(R^2 - x^2)}$, parallel to the y-axis, as shown in Figure 7.8. The Huygens' integral, when combined with (230), gives

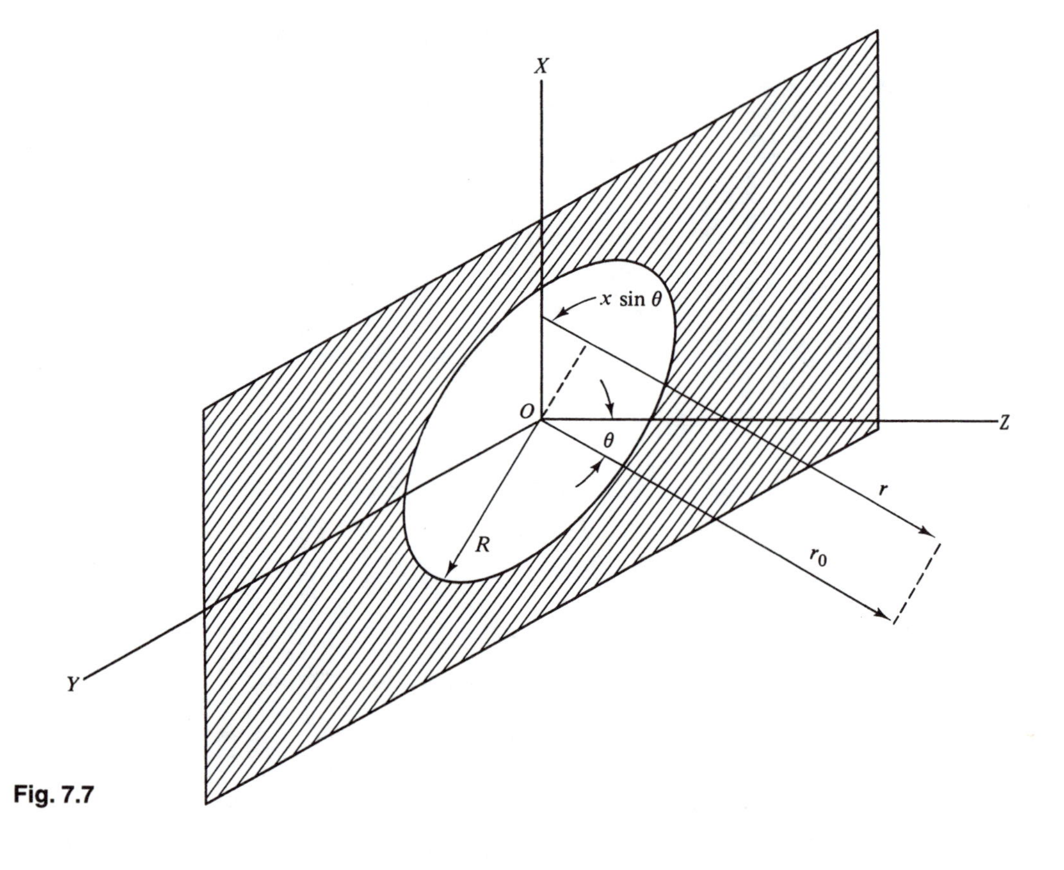

Fig. 7.7

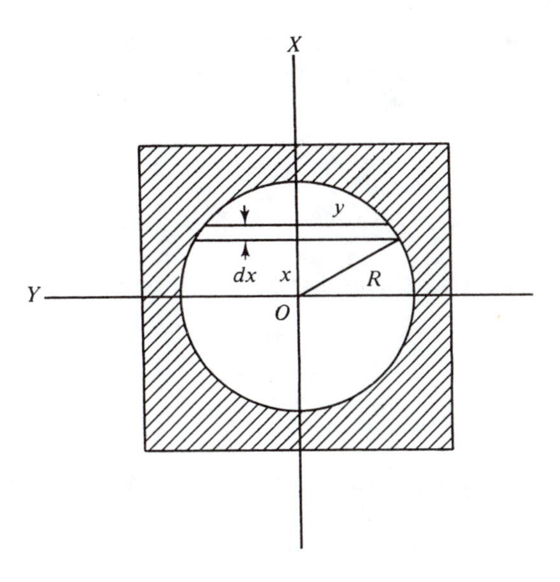

Fig. 7.8

$$A = 2Ce^{ikr_0}\int_{-R}^{R} e^{ikx\sin\theta}\sqrt{R^2 - x^2}\,dx \tag{242}$$

Letting

$$u = x/\mathrm{R},\ \rho = k\mathrm{R}\sin\theta$$

and absorbing the constants into C (as before) converts this to

$$A = \int_{-1}^{1} e^{i\rho u}\sqrt{1 - u^2}\,du \tag{243}$$

The exponential term inside the integral can be written as

$$e^{i\rho u} = \cos\rho u + i\sin\rho u \tag{244}$$

The integral involving the sine vanishes, because the integrand is an odd function. The integrand involving the cosine is an even function, so that we can change the limits from –1, 1 to 0, 1 and put a factor of 2 out front. This makes (243) have the form of a *Bessel function* J_1, defined as

$$J_1(\rho) = \frac{2\rho}{\pi}\int_0^1 (1 - u^2)^{1/2}\cos(\rho u)\,dt \tag{245}$$

Although Bessel functions are normally used in optics books, they are not easy to deal with. We can simplify what follows by using numerical methods. The amplitude at the screen becomes

$$A = \frac{2CJ_1(\rho)}{\rho} \tag{246}$$

Further, when $\theta = 0$, then $\rho = 0$, and the Bessel function definition shows that the quantity $[J_1(\rho)/\rho]$, when evaluated at $\rho = 0$, is equal to 0.5.

PROBLEM 92

Verify (246) from (245).

Then (246) multiplied by its complex conjugate and divided by the quantity just evaluated gives

$$\frac{I}{I_0} = \left[\frac{2J_1(\rho)}{\rho}\right]^2 \tag{247}$$

where the factor of 2 is retained to make the maximum value equal to unity. Problem 93 then deals with the numerical evaluation of the integral in (245) and indicates that we are simply regarding $J_1(\rho)$ as a convenient abbreviation for this integral in (245) rather than requiring any knowledge of the properties of Bessel functions.

PROBLEM 93

(a) Integrate (245) numerically for $\rho = 0, 1, \ldots, 5$ using Simpson's rule (or some convenient algorithm) to reproduce Figure 7.9.

(b) Use this result to find I/I_0, thus verifying Figure 7.10.

(c) Compare Figures 7.6 and 7.10.

The bright central area in the diffraction pattern of Figure 7.10 is known as the *Airy disc* (after a famous British astronomer and mathematician). The first dark ring corresponds to the place where the curve of Figure 7.9 crosses the axis at $\rho = 3.832$. Since

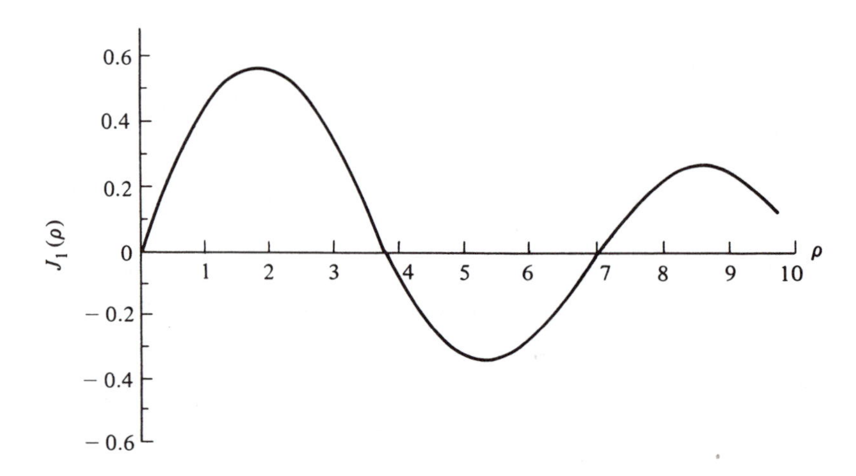

Fig. 7.9

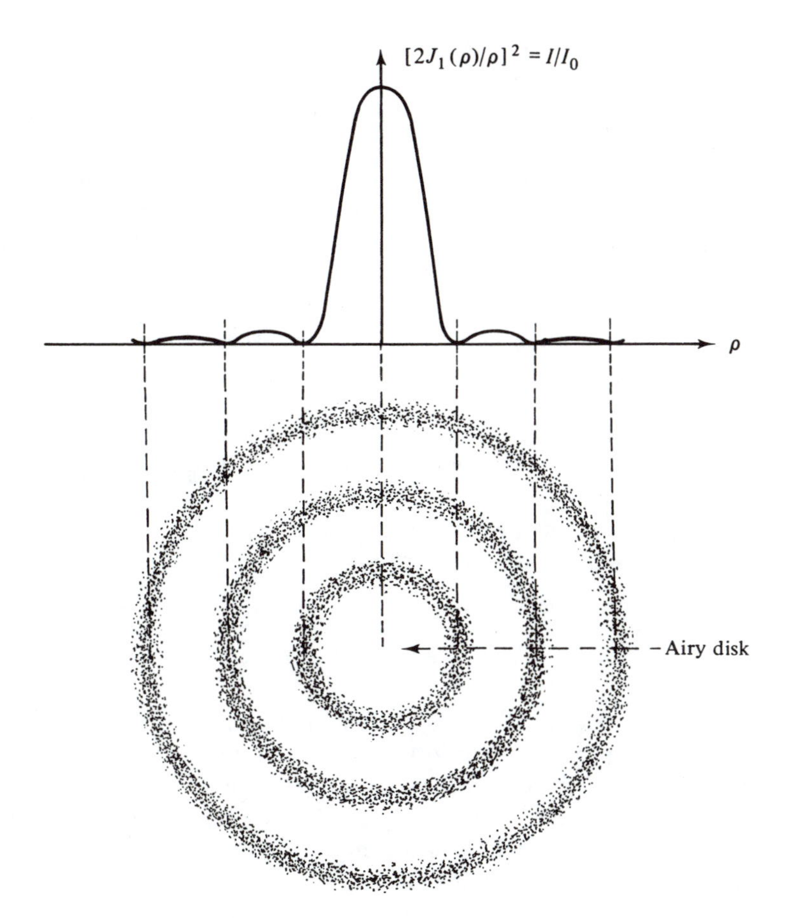

Fig. 7.10

$$\sin \theta = \rho/\mathrm{k}R$$

then the half-angle subtended at the center of the aperture by the diffraction pattern is

$$\theta = \frac{3.83\,\lambda}{2\pi R} = \frac{0.61\,\lambda}{R} = \frac{1.22\,\lambda}{D} \tag{248}$$

where D is the diameter of the aperture. This is a very useful result, as the next section shows.

7.6 LENSES AND DIFFRACTION

The paraxial approximation indicates that all rays leaving an object point will meet at a single point in image space. Since a circular aperture produces an Airy disc rather than a perfect point image, ray optics does not tell the whole story for aberration-free systems, which are said to be *diffraction limited*. Their only defect is the finite size of the image points, which then determines the *resolution*, or ability to distinguish image points. Let two adjacent Airy discs be arranged such that each maximum will fall at the edge of the other one (Figure 7.11); the spacing of the two discs is equal to their radius r. The eye can see two distinct images for such a situation; this resolution condition is called the *Rayleigh criterion*, which describes a diffraction-limited system. Since the Rayleigh criterion depends on the characteristics of the eye, it is not a precise specification, but it can be used to make reasonable estimates of the capability of a system.

PROBLEM 94

(a) Use the Rayleigh criterion to find the diameter of a lens that can just resolve the limbs (two points on the ends of the diameter) of the moon (the diameter D of the moon is 1,080 miles [mi], and it is 240,000 mi away).

(b) Use the Rayleigh criterion to find the resolution at ground level of a space telescope with a lens whose diameter is 2.4 m and orbiting at 275 km.

(c) Confirm (b) with some simple geometry if $f' = 57.6$ m and the film used by the telescope has a resolution of 15 μm.

Consider next the concept of *lens speed*, which is the ability of a lens to collect light from a source. This quantity should obviously depend on the size, or diameter D, of the lens, and it also depends on the focal length f', as indicated by Figure 7.12. This figure shows that the acceptance angle ϕ, which is measure of the ability of the lens to collect light, increases as f' gets shorter so that the speed decreases as f' increases; therefore the acceptance angle ϕ is given by the ratio D/f'. Photographers use the reciprocal of this quantity, calling it the *f-number* or f#. Thus

$$f\# = f'/D$$

However, if a point source of light is moved closer to the lens, then the image will move past the paraxial focus, the angle ϕ will become smaller, and less energy will appear at the image point, so that f# indicates the light-collecting ability only if the source is at a great distance. Photographers are aware of this effect and compensate

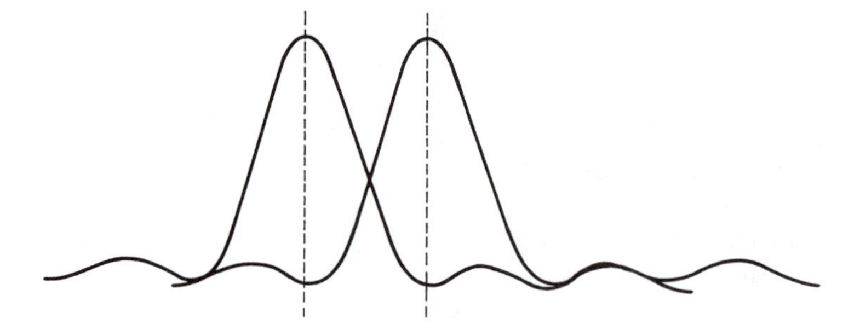

Fig. 7.11

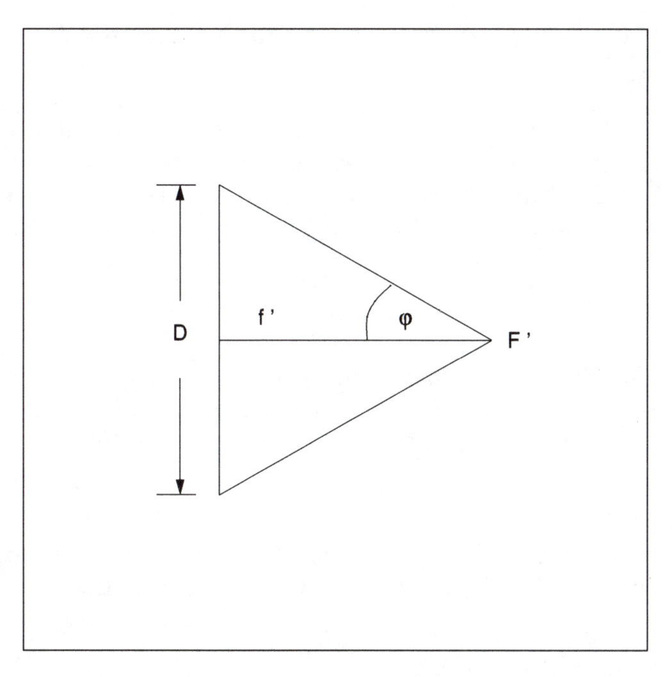

Fig. 7.12

for it. The effective speed of a lens will change if the optical system is immersed in a material medium. The f-number is a quantity that refers to behavior in air; if a lens is in oil, for example, then its focal length f'_O is increased. The focal length f' in air can be obtained from the Gaussian constant as

$$f' = \frac{1}{a} \tag{249}$$

and from Chapter 1

$$f_O' = \frac{n'}{a} = n'f' \tag{250}$$

where n' is the index of the oil. This indicates that the effective focal length has been increased, and the f-number is not the same as in air. For situations where the f-number can change, it is better to use a different way of specifying lens speed. Figure 7.3 shows an Airy disc of radius r and a lens of diameter D. The half-angle θ in radians is approximately r/f' and by (248)

$$D = \frac{1.22\lambda}{r/f'} \tag{251}$$

or

$$r = \frac{1.22\lambda}{D/f'} \tag{252}$$

Using a value of 550 nm for green light, we obtain the diameter d of the Airy disc as

$$d = 1.3 \times 10^{-6} f\#$$

which indicates that, to a reasonable approximation, the f-number of a lens tells us the size of the Airy disc in micrometers. Equation (252) also specifies the resolution available in a diffraction-limited system. The distance between the centers of two Airy discs in Figure 7.13 is the radius r. If we immerse the lens in oil, then (252) becomes

$$r = \frac{0.61\,\lambda}{Rn'/f_O'} \tag{253}$$

using (250) to obtain the altered focal length. The quantity R/f'_O in the denominator will be recognized as a measure of lens speed, using the lens radius rather than the diameter. This quantity is known as the *numerical aperture* NA, so that

$$NA = Rn'/f_O' \tag{254}$$

It is determined by the size of the lens, the focal length in oil, and the index of the oil. For a lens in air, with $n' = 1$ and $f'_O = f'$, this definition shows that

$$NA = 1/(2\,f\#)$$

Combining (253) and (254)

$$r = \frac{0.61\lambda}{NA} \tag{255}$$

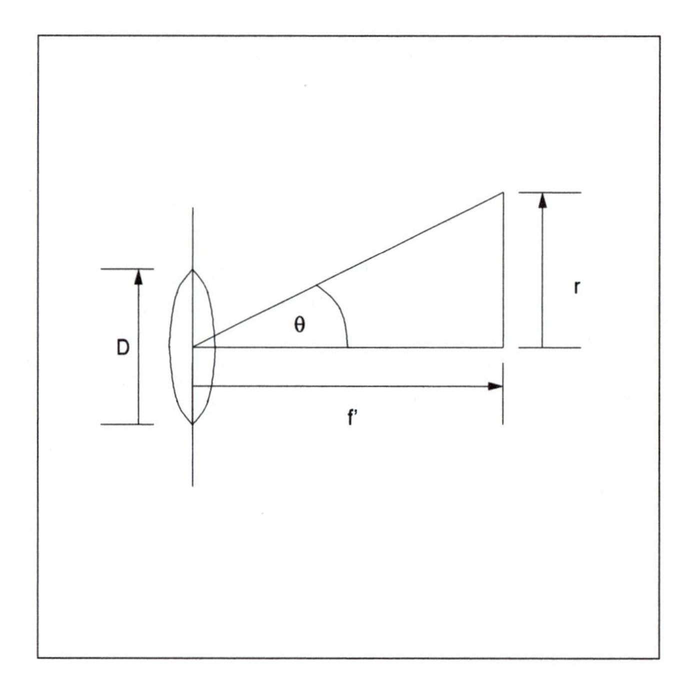

Fig. 7.13

or approximately

$$r = \frac{\lambda}{NA} \tag{256}$$

That is, to find the minimum spacing allowed between the centers of the Airy discs, simply divide the wavelength of the light being used by the numerical aperture of the lens. Now let Figure 7.13 represent a lens immersed in oil. For small angles

$$\sin\phi = \frac{R}{f_O'} \tag{257}$$

so that (266) becomes

$$NA = n'\sin\phi \tag{258}$$

To get an idea of the range of values for the quantity NA, consider the oil immersion microscope objective that is illustrated of Figure 3.10. Cedar oil with an index of 1.517 results in a critical angle of 71°, so that

$$NA = 1.517 \sin 70° = 1.43$$

It is possible to use other media that give values up to 1.6. Less elaborate microscope objectives will have numerical apertures well below 1.0.

PROBLEM 95

By adding the image spread due to diffraction and to chromatic dispersion and then differentiating, show that the combined spread will be a minimum when

$$R = \sqrt{\frac{1.22\lambda f'}{V}} \tag{259}$$

PROBLEM 96

It is possible to get an approximate version of (248) by using the results for a rectangular slit. Let such a slit have a length equal to the diameter of a circular opening and require that the areas be the same. Show that the first minimum in the diffraction pattern of the rectangular slit is specified by the relation

$$\theta = \frac{1.27\lambda}{D} \tag{260}$$

which represents a deviation of only 5% from the exact calculation.

PROBLEM 97

Consider a rectangular sheet of transparent material that is infinite in both the x and the y directions and whose light transmission properties are specified by the Gaussian function

$$f(x) = Be^{-x^2/b^2} \tag{261}$$

where B and b are constants. Show that the diffraction pattern is also Gaussian and does not display the secondary maxima of the rectangular and circular apertures. This indicates that, when an opening does not terminate abruptly and there is no sharply defined edge, the diffraction pattern is similar to the incident wave.

PROBLEM 98

A metal plate has two rectangular slits as shown in Figure 7.14. Show that

$$\frac{I}{I_0} = \left(\frac{\sin\alpha}{\alpha}\right)^2 \cos^2\gamma \tag{262}$$

where

$$\alpha = {}^1\!/\!_2 kw \sin\theta, \ \gamma = {}^1\!/\!_2 kd \sin\theta$$

A plot of this equation is shown in Figure 7.15.

Figure 7.15 indicates that there are two factors that contribute to the diffraction pattern for this pair of openings—one comes from the original single opening and the other from its repetition at a distance d. The solid curve, which represents I/I_0 vs. θ, is the overall pattern. The dotted curve is the envelope of the solid curve and has the

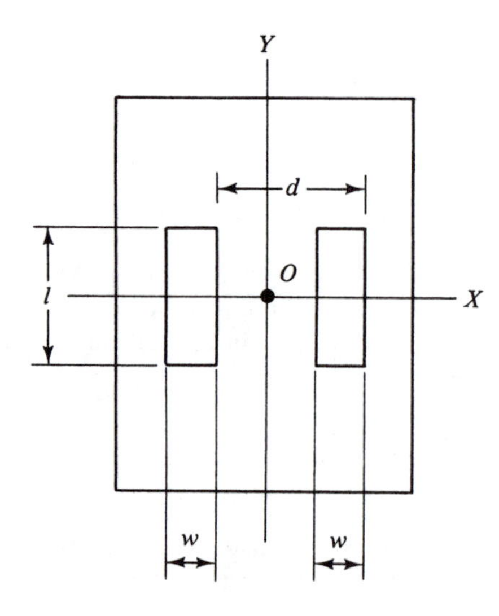

Fig. 7.14

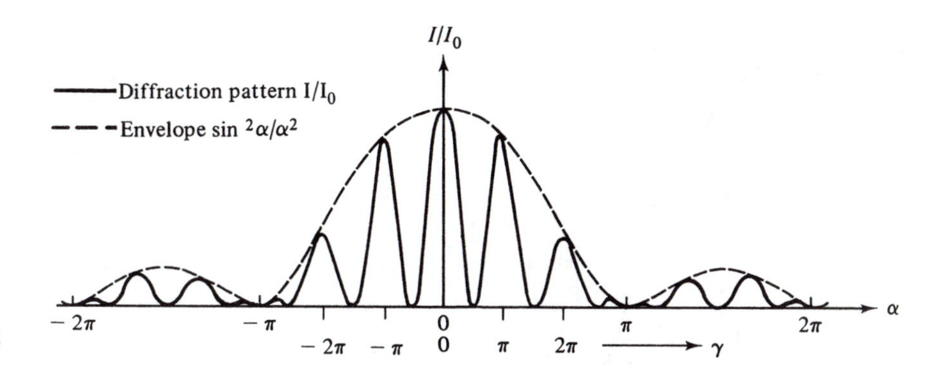

Fig. 7.15

form of Figure 7.6. The function (sin α)/α is known as the *sampling function* $Sa(\alpha)$; in optics, it is usually written as $(\sin \pi x)/\pi x$, where $x = w \sin\theta$. It is denoted sinc x and called the *sinc function*.

7.7 APODIZATION

The diffraction patterns that have been determined for rectangular and square openings have bright central sections containing most of the transmitted energy, while they have rather feeble secondary sections. However, for applications such as the observation of very dim stars, it would be nice to transfer some of the energy outside the Airy disk of the telescope into the central section. This process is called *apodization*, from the Greek, meaning "without feet." We have already seen in Problem 97 that a transparent sheet can alter a diffraction pattern. Let's place a sheet whose transmission properties vary as $\cos(\pi x/w)$ across a rectangular aperture. The diffraction integral (232) then becomes

$$A = Ce^{ikr_0}l\int_{-w/2}^{w/2} \cos\left(\frac{\pi x}{w}\right) e^{ikx\sin\theta} dx \tag{263}$$

PROBLEM 99

(a) Evaluate this integral to obtain

$$\frac{I}{I_{0A}} = \left(\frac{\pi^2 \cos\alpha}{\pi^2 - 4\alpha^2}\right)^2 \tag{264}$$

where I_{0A} is the value of the intensity in the apodized opening when $\alpha = 0$.

(b) Show that

$$I_{0A} = 1.625 I_0 \tag{265}$$

where I_0 refers to the original opening.

(c) Verify Figure 7.16 numerically.

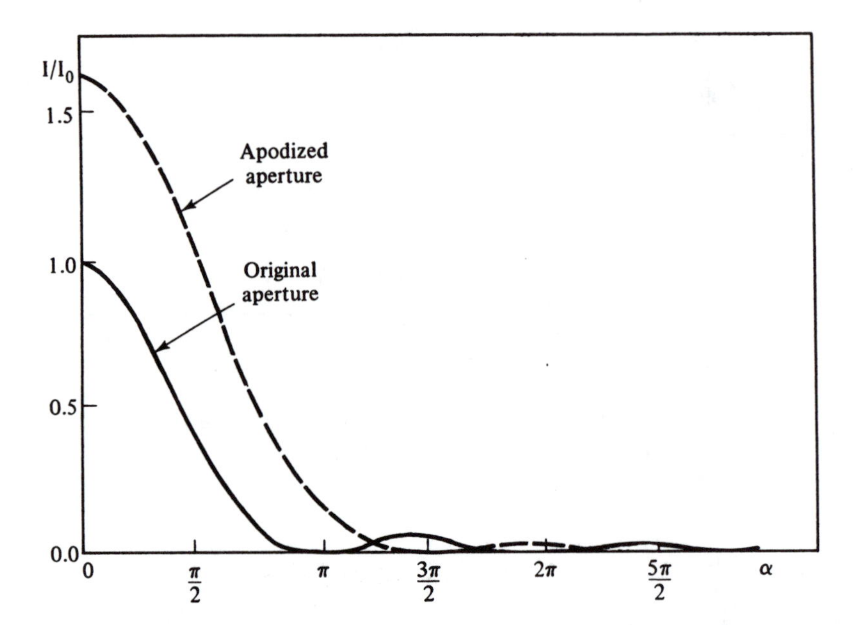

Fig. 7.16

The apodizing sheet has thus raised the height and width of the central maximum by shifting energy from the secondary region. We note, however, that the only way to get all the energy into a central band is to use the infinite Gaussian material mentioned in Problem 97.

7.8 THE DIFFRACTION GRATING

Having determined the diffraction pattern for a pair of identical rectangular apertures, let's see what happens when we have a total of N such openings with a uniform spacing (Figure 7.17). The diffraction integral is then the sum of the individual contributions from the N openings, or

$$A = \int_0^w e^{ikx \sin\theta} dx + \int_d^{d+w} e^{ikx \sin\theta} dx + \ldots\ldots + \int_{(N-1)d}^{(N-1)d+w} e^{ikw \sin\theta} dx \tag{266}$$

Evaluating the integrals, we obtain

$$A = \frac{e^{ikw \sin\theta} - 1}{ik \sin\theta}\left(1 + e^{ikd \sin\theta} + \ldots . + e^{ik(N-1)d \sin\theta}\right) \tag{267}$$

Fig. 7.17

The terms in parentheses on the right form a geometric progression whose sum S is

$$S = \frac{a(1 - r^N)}{1 - r} \tag{268}$$

where r is the ratio, a is the first term, and N is the number of terms. Using α and γ as previously introduced in connection with equation (262), the amplitude of the diffraction pattern is

$$A = \left(\frac{e^{2i\alpha} - 1}{2i\alpha/w}\right)\left(\frac{1 - e^{2i\gamma N}}{1 - e^{2i\gamma}}\right) \tag{269}$$

But

$$e^{2i\alpha} - 1 = (e^{i\alpha} - e^{-i\alpha})e^{i\alpha} = 2e^{i\alpha}i \sin\alpha \tag{270}$$

with similar expressions for the exponentials involving γ. When these are combined with (269) and the result is multiplied by its complex conjugate, we obtain

$$\frac{I}{I_0} = \left(\frac{\sin\alpha}{\alpha}\right)^2\left(\frac{\sin N\gamma}{N\sin\gamma}\right)^2 \tag{271}$$

The quantity N^2 is deliberately incorporated into the denominator of the second term so that, when θ approaches zero and $\sin^2 N\gamma$ can be approximated by $(N\gamma)^2$, the entire right-hand side will reduce to unity and I will approach I_0, as it should. This equation expresses the diffraction pattern as the product of two terms. The first of these terms, called the *shape factor*, describes the contribution of a rectangular opening; we obtained the same result for both the single and the double slit. The second term modifies the pattern to take into account the fact that there are N identical slits; this is the *grating factor*. This term is a generalization of what we found when $N = 2$, for

$$\sin 2\gamma/2\sin\gamma = \cos\gamma$$

A plot of the intensity for two values N = 5 and N = 20 is shown in Figure 7.18. The shape factor is the envelope, as in Figure 7.15. The central maximum of the single opening has become a series of closely spaced sharp lines as N gets larger. These maxima occur for values of γ that make the term $\sin\gamma$ in the denominator of the grating factor vanish; that is, for

$$\gamma = m\pi, \qquad m = \pm1, \pm2, \ldots \tag{272}$$

Combining this with the definition of γ, we obtain

$$\tfrac{1}{2}kd\sin\theta_m = m\pi$$

or

$$d\sin\theta_m = m\lambda \tag{273}$$

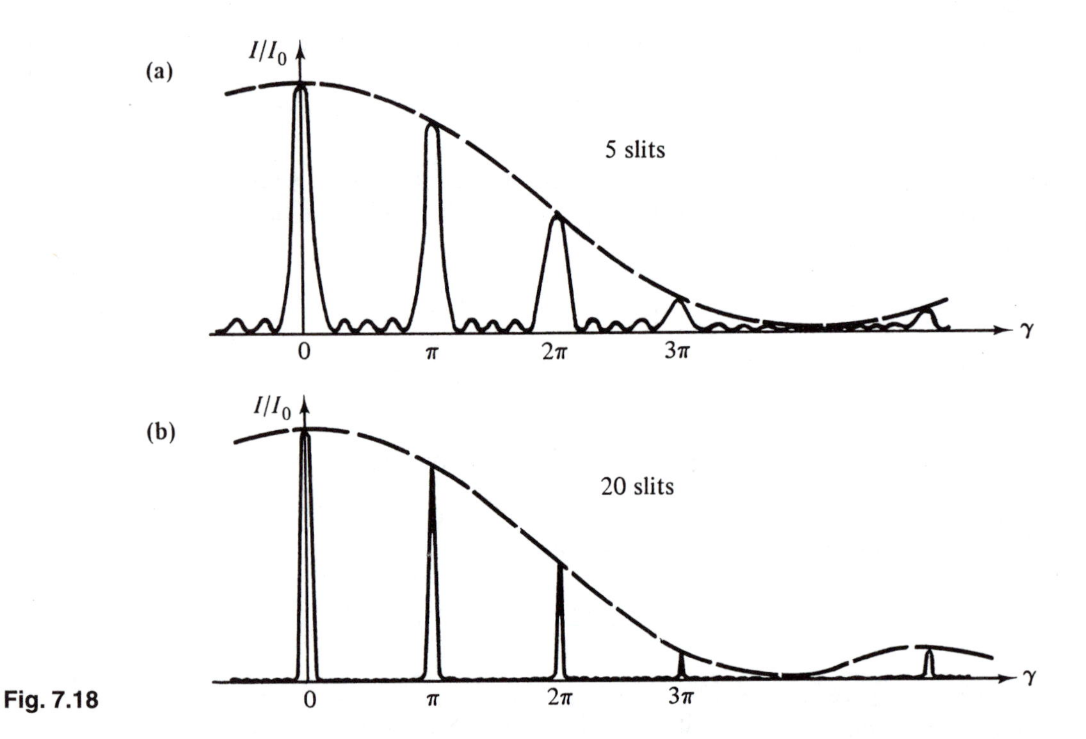

Fig. 7.18

where θ_m denotes the angles with respect to the normal. Equation (273) will be recognized as a form of *Bragg's law*, which governs the diffraction of light and of electrons by single crystals. The connection should not be surprising, since a crystal and a diffraction grating are both structures with a regular periodicity. Keep in mind that these results are valid only for small angles, since we have used a very primitive kind of diffraction integral.

Figure 7.19 shows rays for which m = 0, +1, or −1. On a screen at some distance away, they will produce three sharp lines, the middle one being slightly brighter than the others. The complete set of rays is called the *spectrum* of the grating, and the value of m is the *order* of the individual members of this spectrum. The separation of the orders shown in the figure has a practical use, as will be considered next.

7.9 Spatial Filtering

Place a lens of focal length f' at a reasonable distance from a grating. Since we have assumed small angles, the rays of order 0 will meet at the paraxial focus F′. Those of order +1 will meet in the paraxial plane at a short distance above the axis, and those for $m = -1$ will meet at the same distance below the axis. This situation is illustrated in Figure 7.20.

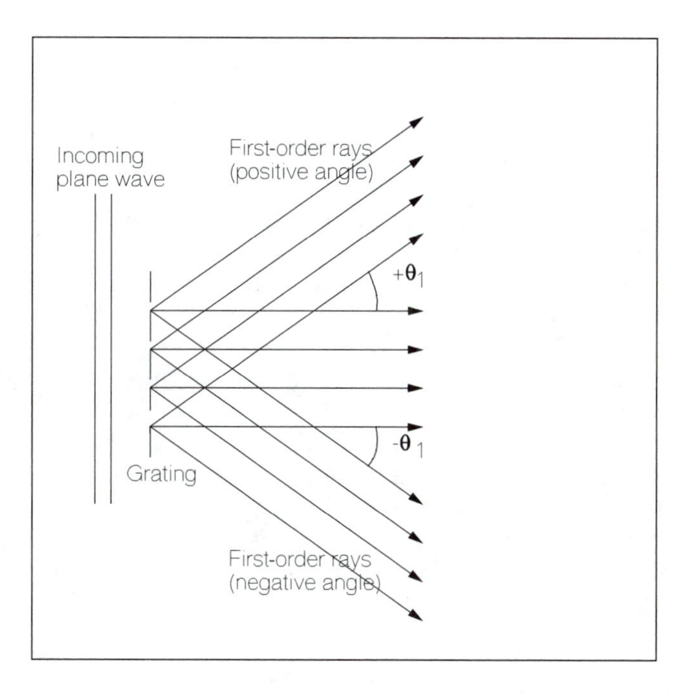

Fig. 7.19

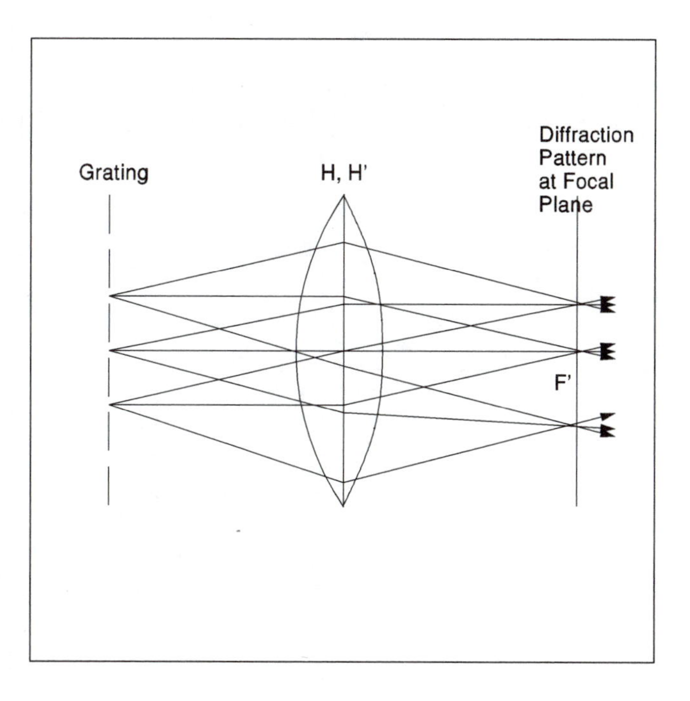

Fig. 7.20

PROBLEM 100

Show that the half-width x' of the central line is $x' = f\lambda/Nd$ and that the first-order line is a distance $f\lambda/d$ from the axis.

For a thin lens, all the refraction occurs at the two coincident unit planes, the zero-order rays meet at F′, and the rays for m = +1 and m = −1 meet as shown. This series of sharp lines is the diffraction pattern. If these rays extended to the right (Figure 7.21), they form the real image of the diffraction grating. Next, we determine the nature of this image. When a plane wave strikes the grating, a series of waves emerges. The wave for $m = 0$ travels parallel to the axis and is brought to a focus at the central image point. The wave for $m = +1$ is deflected upward by a small angle and produces the next bright region in the paraxial image plane, and so on for all the other orders. The waves for $m = 1$ or -1 take slightly longer to reach this plane than the wave for $m = 0$; they are out of phase with the central wave. This can be handled analytically by incorporating a phase lag ϕ_m into the exponent of the solution (228) to the wave equation. We can therefore express the amplitude of the image of any one of the slits in the grating as

$$A = \sum_{m=-\infty}^{\infty} A_m e^{i\phi_m} \tag{274}$$

The exponential term can be written as

$$e^{i\phi_m} = \cos\phi_m + i\sin\phi_m \tag{275}$$

and, for every term in the summation except the one for m = 0, the terms in $\cos\phi_m$ will add and those in $\sin\phi_m$ will cancel. Using the shape factor of equation (233), this becomes

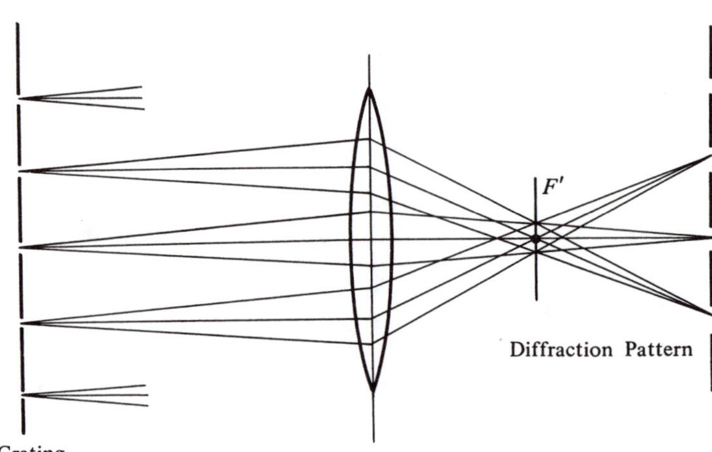

Fig. 7.21

$$\frac{A}{C'} = \sum_{m=-\infty}^{\infty} \frac{\sin \alpha_m}{\alpha_m} \cos \phi_m \tag{276}$$

with the factor of 2 being absorbed in the constant and where

$$\alpha = (kw/2) \sin \theta_m \tag{277}$$

Equation (276) will become

$$\frac{A}{C'} = 1 + 2 \sum_{m=1}^{\infty} \frac{\sin \alpha_m}{\alpha_m} \cos \phi_m \tag{278}$$

since $\phi_0 = 0$ and $(\sin x)/x = 1$ for $x = 0$. To specify ϕ_m, we use the fact that, for a slit located at a distance x from the axis, the quantity $x \sin \theta_m$ ($x\theta_m$ for small angles) expresses the difference in path length for a ray of order m leaving this slit and a ray leaving the center (the geometry is exactly like that of Figure 7.5). Hence, $i\phi_m$ is equivalent to $ik(x\theta_m)$. But by (273)

$$\theta_m = \frac{m\lambda}{d} \tag{279}$$

so that

$$\phi_m = \frac{2\pi m x}{d} \tag{280}$$

The image plane location is given by Gauss' law, which we use in the form

$$\frac{s'}{s} = \frac{s' - f'}{f'} = \frac{x'}{x} \tag{281}$$

Solving for x, (280) becomes

$$\phi_m = \frac{2\pi m x' f'}{(s' - f')d} \tag{282}$$

We also need α_m, and, by (277) and (279)

$$\alpha_m = \frac{m w \pi}{d} \tag{283}$$

Substituting into (278) and squaring

$$\frac{I}{I_0} = \left[1 + 2 \sum_{m=1}^{M} \frac{\sin(mw\pi/d)}{mw\pi/d} \cos \frac{2\pi m x' f'}{(s' - f')d}\right]^2 \tag{284}$$

where M designates the number of terms to be used in the summation.

This expression comes from A. Garrard (*American Journal of Physics* 31, 723 [1963]), and he applies it to a grating with 5,000 lines per cm, the opaque regions being 50% larger than the slits. Then d = 0.0002 cm and w = 0.8 μm. The lens has a focal length of 1 cm, and s = 1.2 cm so that s' = 6.0 cm. Since the magnification s'/s is 5.0, the slit images are 4.0 μm wide. These numbers give

$$\alpha_m = 1.256\, m, \ \phi_m = 6283.1\, mx'$$

and (284) becomes

$$\frac{I}{I_0} = \left[1 + 2\sum_{m=1}^{M} \frac{\sin(1.256m)}{1.256m}\cos(6283.1\, mx')\right]^2 \tag{285}$$

PROBLEM 101

Use (285) to find the relative intensity I/I_0 as a function of position x' in the image for M = 1, 2, 4, 6, and 11, verifying Figure 7.22a–e.

Equation (285) and the plots of Figure 7.22 provide some insight into how optical systems function. Suppose that the diffraction pattern consisted of only the rays for m = 0. This order is what gave the first term on the right of (285) and it would reduce to

$$I/I_0 = 1$$

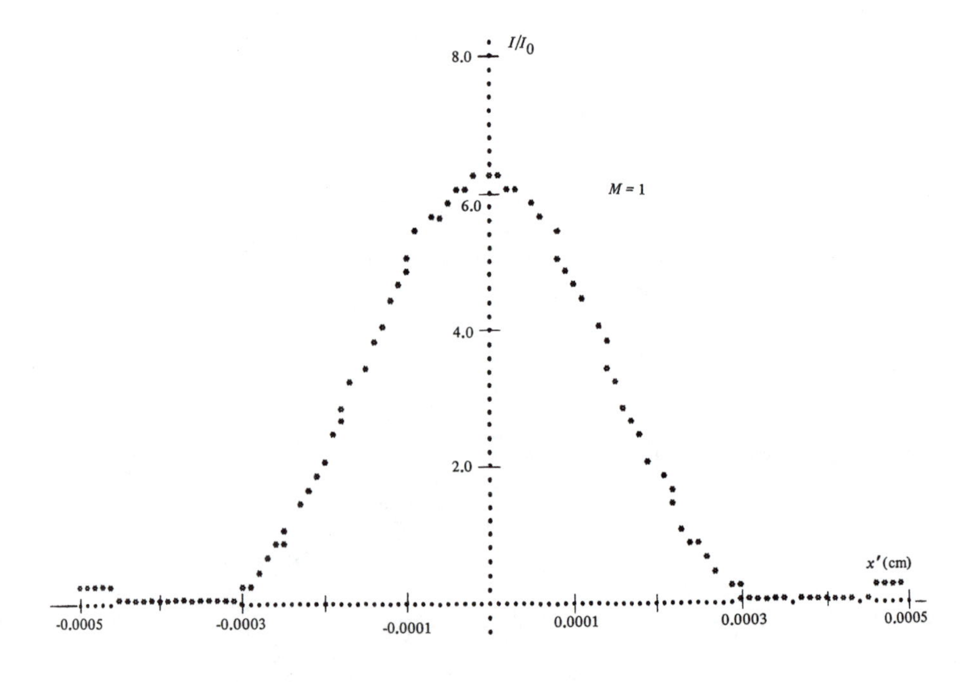

Fig. 7.22a

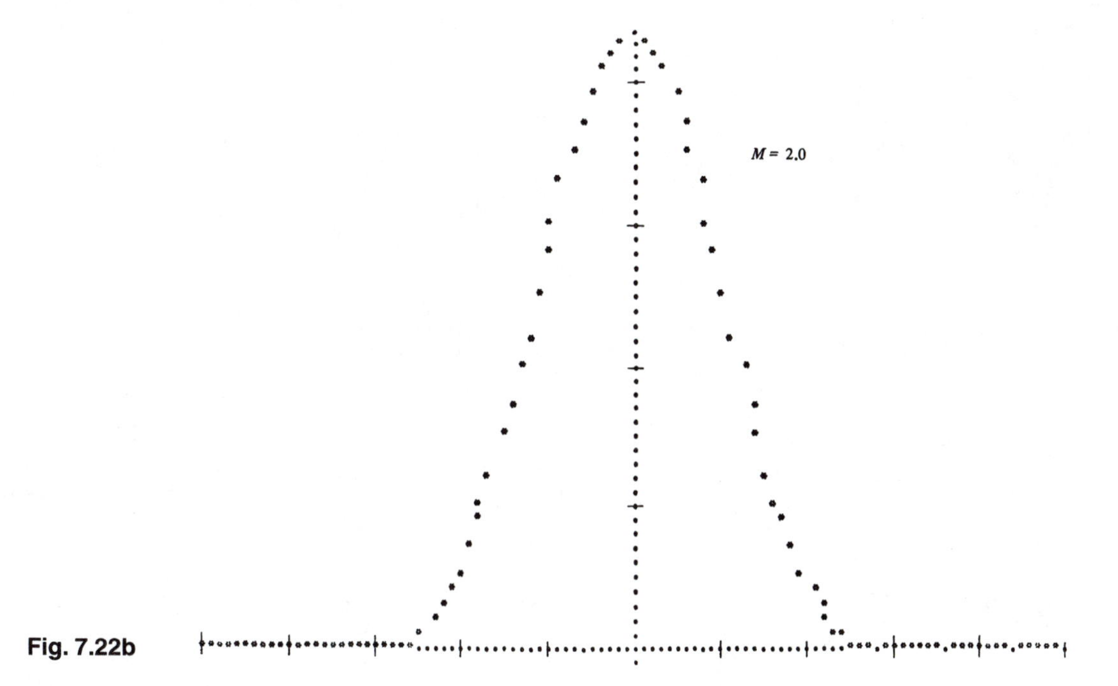

Fig. 7.22b

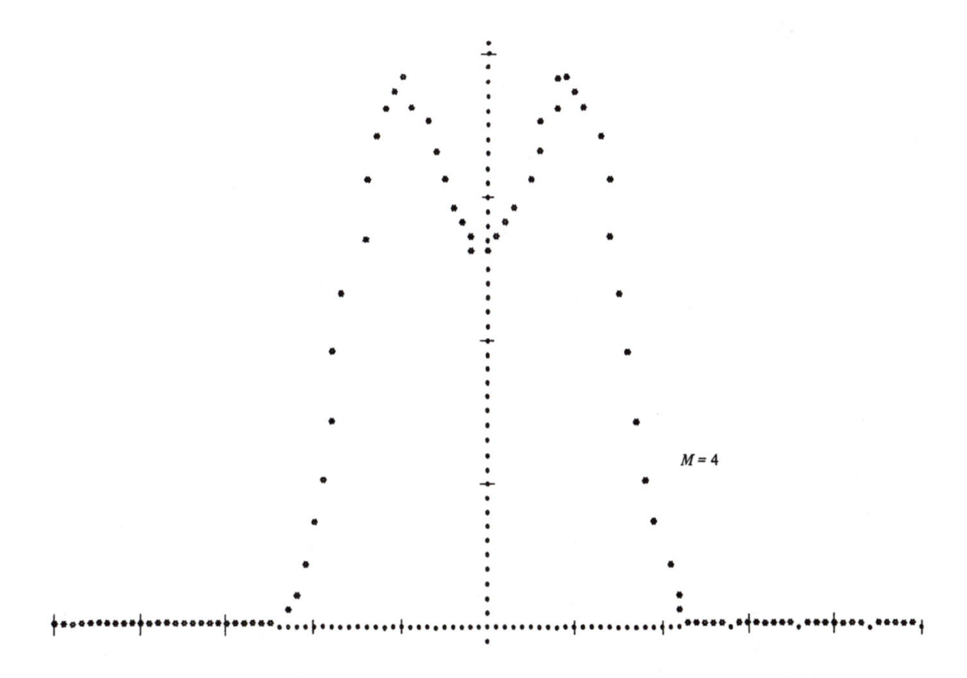

Fig. 7.22c

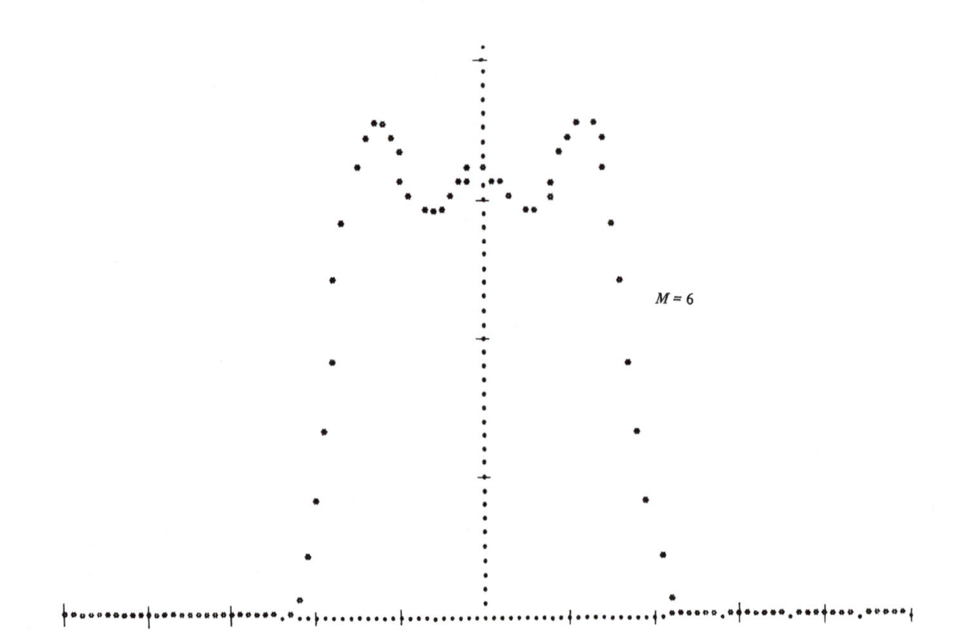

Fig. 7.22d

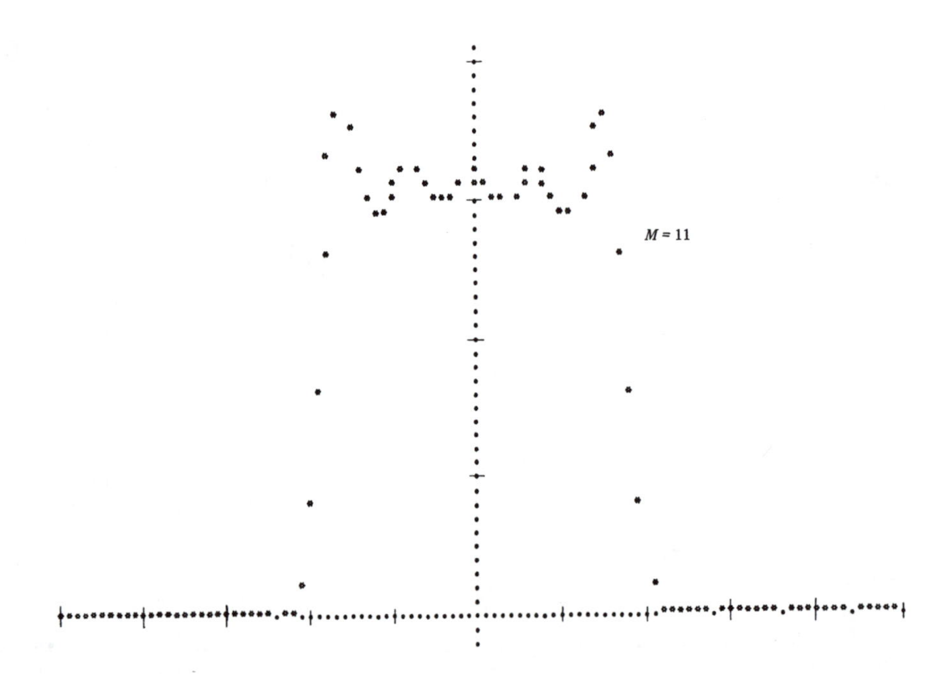

Fig. 7.22e

so that the image would be a completely and uniformly illuminated screen with no detail about the size or shape of the slit. Now add the two first-order terms to the zero-order term. The result, Figure 7.22a, indicates that the image is recognizable as an opening of about 6 μm in width. This image is quite different from the sharp and even image called for by geometric optics. However, as we add orders, the image structure improves, and Figure 7.22e displays an image of the predicted width and with a shape coming closer to a rectangular structure. What we have just expressed here is the *Abbe theory of the microscope*. Abbe realized that, in order to get any kind of recognizable image, it was necessary to transmit at least the zero-order and first-order rays from the object, and the more higher orders that were used, the better the resolution. This represents a basic advantage of the oil immersion objective—the very large angle at the edge of the cover glass in Figure 7.22 means that very high orders can get to the objective. Of course, the microscope user is not looking at diffraction gratings, but all transparent objects diffract light, and the situation is much more complicated than the ideal arrangement considered here.

Figure 7.20 reveals what can happen if the higher orders are blocked. Suppose we place at the focal plane a metal plate with an aperture whose radius is slightly smaller than the distance between the first-order rays. Only the zero-order rays can get through; thus we have created a *high-pass filter*. For the left-hand side of Figure 7.22b, the lens can receive only the lowest orders, and it is acting as such a filter. This is a situation for which filtering degrades the image, but there are applications for which it is deliberately introduced. Figure 7.23a shows a self-portrait by the Minnesota artist Karl E. Bethke, as reproduced in a paper by R. A. Phillips, *American Journal of Physics,* 37, 536 (1969); his painting consists of regularly spaced squares on canvas. Using a pinhole as a filter, the higher-order components are removed and the sharp edges of the squares have been removed (Figure 7.23b). The result is not terribly good because the higher orders are responsible for detail, much of which has been

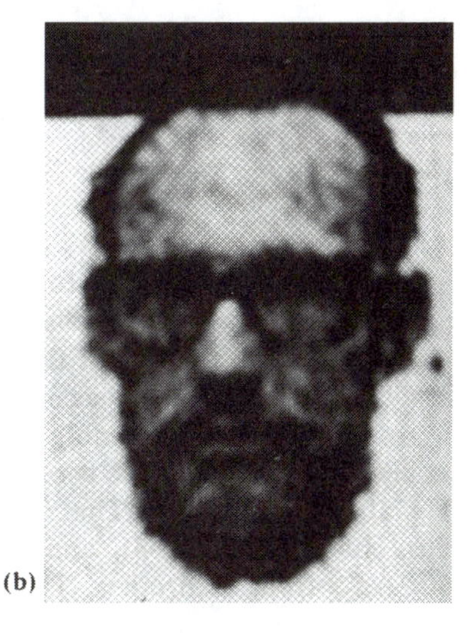

Fig. 7.23 (a) (b)

lost. This procedure is used by NASA to process satellite photographs. They are assembled in strips, and the regularly spaced edges can be eliminated by high-pass filtering.

A simple experiment that demonstrates several of these ideas was described by K. J. van Camp (*American Journal of Physics* 37, 105 [1969]). Figure 7.24a shows a one-celled organism called a *diatom*. Because of its regular structure and small size, it has been used as a test of the resolution of microscopes. The experiment involved fastening a V-shaped diaphragm (Figure 7.24c) to the objective; the exact position with respect to the focal plane on the image side is not crucial. When the slots in the diaphragm are lined up with the specimen, the microscope functions normally. If the diaphragm is rotated by 90°, the opaque sections block the higher orders and the regular structure vanishes (Figure 7.24b). This experiment shows that image formation involves diffraction; it is an elementary example of spatial filtering.

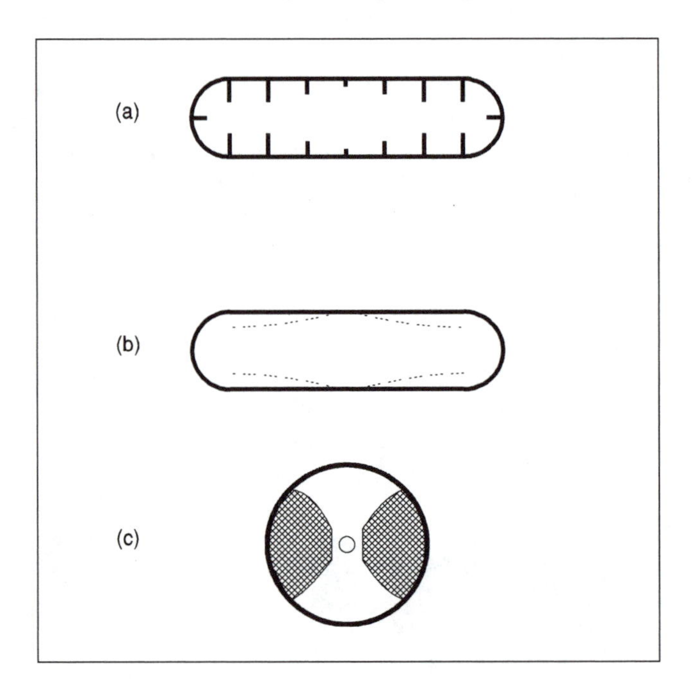

Fig. 7.24

CHAPTER 8

Fourier Analysis

8.1 Fourier Series

To start very simply, Figure 8.1 shows a plot defined by the conditions

$$y = \pi/4 \text{ for } -\pi/2 \leq x \leq \pi/2,\ y = -\pi/4 \text{ otherwise}$$

This rectangular function has the disadvantage of being discontinuous at $x = \pi/2$ and $x = -\pi/2$, and we want to replace it with a function that is everywhere continuous. Start with a half-cycle of a cosine curve as a crude approximation. Figure 8.2a at the top shows $y = \cos x$ and it is repeated at the bottom, where we see that it is not a very good match. But if we subtract the term (1/3)cos $3x$, shown in Figure 8.2b, we obtain the slight improvement of Figure 8.2d. Adding (1/5) cos ($5x$) (Figure 8.2c) gives a further improvement, Figure 8.2e. We now have

$$y = (1) \cos x - (1/3) \cos 3x + (1/5) \cos 5x$$

which is a *Fourier series*. We would expect that the additional terms –(1/7) cos $7x$, (1/9) cos $9x$, and so on, would continue to improve the approximation. It turns out, however, that the series overshoots the square wave at the edges ($x = \pm\pi/2$), and failure of the Fourier series to give a good match is known as the *Gibbs phenomenon*.

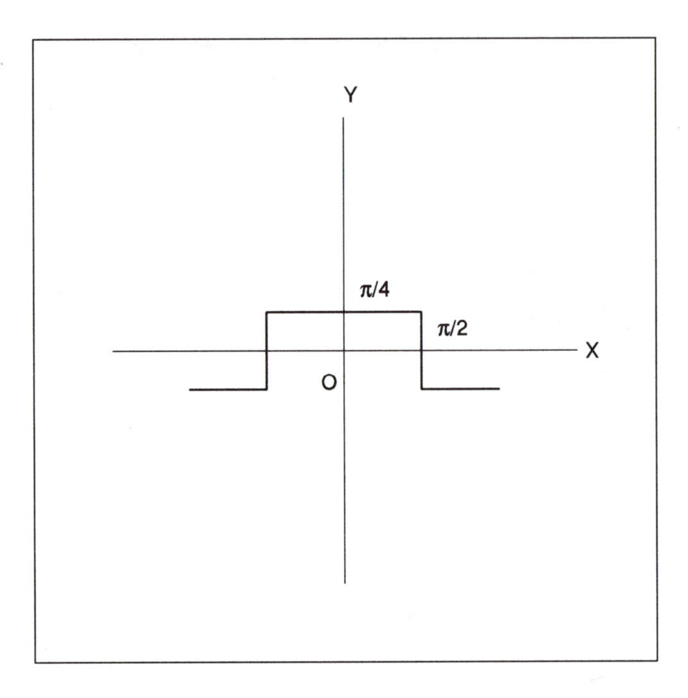

Fig. 8.1

PROBLEM 102

(a) Write a program to verify the curves shown in Figure 8.2.

(b) Extend this program to a series containing 25 terms and explain why the Gibbs phenomenon corresponds to a 9% overshoot.

Although the Gibbs phenomenon is important for the design of spatial filters, it will not affect our use of the Fourier theory. However, another problem that does appear is that Figure 8.2 shows only a part of each of the cosine waves. Since the cosine is periodic in both directions, our attempt to convert a single rectangular pulse into a continuous function leads to a periodic train of square waves; we will consider this matter shortly.

Although we have used cosines to construct a Fourier series, a rectangular structure with its left-hand edge on the y-axis would be better matched by sines, and a more general Fourier series involves sines and cosines, being of the form

$$y = \sum_{n=0}^{\infty} (a_n \cos nx + b_n \sin nx) \tag{286}$$

Sines and cosines have a period of 2π. We shall want to use expressions valid for periods of any length, so that this Fourier series has to be modified by replacing nx with nkx, where k is the propagation constant previously defined as

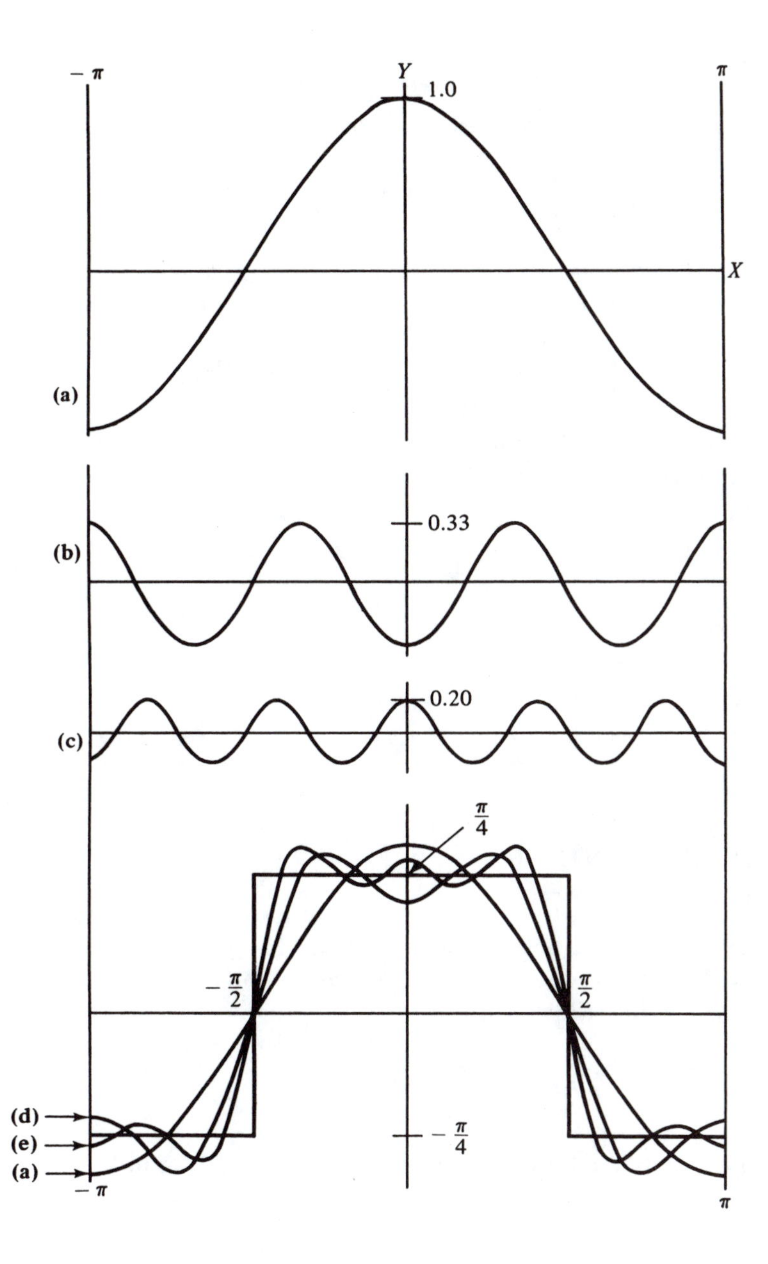

Fig. 8.2

(287)

Another change involves recognizing that exponential functions are more convenient to deal with than sines and cosines, the two types being related by the Euler identity

$$e^{inkx} = \cos nkx + i\sin nkx \tag{288}$$

Then the Fourier series that is equivalent to (286) will have the form

$$y = f(x) = \sum_{n=-\infty}^{\infty} c_n\, e^{inkx} \tag{289}$$

where c_n is complex. We can obtain an explicit expression for c_n if we multiply (301) by e^{-imkx}, where m is an integer, and integrate from $-\lambda/2$ to $\lambda/2$ to obtain

$$\int_{-\lambda/2}^{\lambda/2} f(x)e^{-imkx}dx = \sum c_n \int_{-\lambda/2}^{\lambda/2} e^{i(n-m)kx}dx \tag{290}$$

The term in the sum for which $m = n$ has a value of $c_n\lambda$ when integrated, so that the coefficients are given by the formula

$$c_n = \frac{1}{\lambda}\int_{-\lambda/2}^{\lambda/2} f(x)e^{-iknx}dx \tag{291}$$

PROBLEM 103

Find the Fourier series for a periodic set of pulses like those of Figure 8.1.

(*Hint:* Let $n = 0, 1, -1$, and so on, and evaluate each integral separately.)

8.2 THE FOURIER INTEGRAL

We now consider how to represent the single square pulse of Figure 8.1 as a Fourier series. To accomplish this, let us specify its wavelength as infinite in length; it is a pulse train with only one member. To avoid confusion, relabel the variable x in (291) as x' and substitute into (289), obtaining

$$f(x) = \sum_{n=-\infty}^{\infty} \frac{1}{\lambda}\int_{-\lambda/2}^{\lambda/2} f(x')e^{-iknx'}dx'\; e^{inkx} \tag{292}$$

Then multiply and divide by k; this gives a term $1/k\lambda$ or $1/2\pi$ in front of the integral, so that

$$f(x) = \sum_{n=-\infty}^{\infty} \frac{1}{2\pi}\int_{-\lambda/2}^{\lambda/2} f(x')e^{-iknx'}dx'\; e^{inkx}k \tag{293}$$

If the wavelength λ becomes very large, then k must become infinitesimal, and we write k at the end of (293) as dk. The integer n can become large as well, so that we require the product nk to remain finite and designate it simply as k. Since nk can take on a large number of closely spaced values, the summation can be replaced by an integration. Dropping the primes and rearranging, the result is

$$f(x) = \frac{1}{2\pi}\int_{-\infty}^{\infty}\left[\int_{-\infty}^{\infty} f(x)e^{-ikx}dx\right]e^{ikx}dk \tag{294}$$

This is known as the *Fourier integral theorem*, and it leads to an important concept. Define the *Fourier transform* of $f(x)$ as

$$FT[f(x)] = F(k) = \int_{-\infty}^{\infty} f(x)e^{-ikx}dx \tag{295}$$

Using this definition, (294) becomes

$$f(x) = \frac{1}{2\pi}\int_{-\infty}^{\infty} F(k)e^{ikx}dk \tag{296}$$

and this should be compared with (289). In place of a sum over a discrete set of exponentials and the associated coefficients, there is an integral involving a continuous set of coefficients. Thus, what was a Fourier series for a periodic pulse train becomes an integral for a single pulse, and this pulse involves all possible values of k or of λ. Furthermore, since we can get $F(k)$ from $f(x)$ by what we call a direct Fourier transform (equation [295]), we call (296) an *inverse Fourier transform* with the definition

$$FT^{-1}[F(k)] = \frac{1}{2\pi}\int_{-\infty}^{\infty} F(k)e^{ikx}dk = f(x) = FT^{-1}FT[f(x)] \tag{297}$$

The two functions $f(x)$ and $F(k)$ form a *Fourier transform pair.*

8.3 THE TRANSFER FUNCTION

Let's look to circuit theory for help in understanding a concept that is widely used in optics. Suppose we have a resistance of 4 ohms in series with an inductance of 2 henrys. To this circuit apply a voltage $v(t)$ of the form

$$v(t) = \begin{bmatrix} 0 & \textit{for } t<0 \\ 10e^{-t} & \textit{for } t\geq 0 \end{bmatrix} \tag{298}$$

The total voltage is the sum of the potentials across the resistor and the inductor. Designating the current as $i(t)$, which avoids confusion with $i = \sqrt{(-1)}$, we then have

$$v(t) = 4i(t) + 2\frac{d}{dt}i(t) \tag{299}$$

where the first term on the right is a consequence of Ohm's law and the second term expresses the fact that the voltage across the inductance is determined by the rate of change of the current it carries. Starting at time $t = 0$, the differential equation governing the behavior of the circuit will be

$$2\frac{di(t)}{dt} + 4i(t) = 10e^{-t} \tag{300}$$

This can be solved with a well-known procedure. Write it as

$$\frac{di(t)}{dt} + 2i(t) = 5e^{-t} \tag{301}$$

and temporarily let the right-hand side be zero. The resulting equation can be solved by direct integration to obtain

$$i(t) = A\, e^{-2t}$$

where A is a constant to be calculated. We now try as a solution of the original equation the sum of this result and the term in (301) that was dropped to obtain

$$i(t) = A\, e^{-2t} + 5\, e^{-t}$$

Then when we use the initial condition that $i(t) = 0$ at $t = 0$, it follows that $A = -5$, and the final solution becomes

$$i(t) = 5[\exp(-t) - \exp(-2t)]$$

as can be verified by substitution.

Another way of solving such equations, and one that will determine the arbitrary constants automatically, is to take the Fourier transform of each side of (300). Since the variable is now time rather than the coordinate x, we change (295) to

$$F(\omega) = \int_{-\infty}^{\infty} f(t)e^{-i\omega t}dt \tag{302}$$

where ω is the angular frequency associated with the voltage or current, and the exponential term is the time-dependent part of (228). To find the Fourier transform of the derivative $di(t)/dt$, integrate by parts to obtain

$$\int_{-\infty}^{\infty} \frac{di(t)}{dt} e^{-i\omega t}dt = i(t)e^{-i\omega t}\Big|_{-\infty}^{\infty} + i\omega\int_{-\infty}^{\infty} i(t)e^{-i\omega t}dt \tag{303}$$

The first term on the right will vanish when this result is applied to the circuit, since the voltage is not turned on when t is negative and, when t is very large, the exponential goes to zero. This leaves

$$FT\left[\frac{d}{dt}i(t)\right] = i\omega FT[i(t)] = i\omega I(\omega) \tag{304}$$

The Fourier transform of the right-hand side of (301) is

$$V(\omega) = \int_{0}^{\infty}(5e^{-t})e^{-i\omega t}dt = \left[5\frac{e^{-(1+i\omega)t}}{-(1+i\omega t)}\right]_0^{\infty} = 5\left[\frac{1}{1 + i\omega}\right] \tag{305}$$

so that the Fourier transform of the complete equation becomes

$$[(i\omega) + 2]I(\omega) = \frac{5}{1 + i\omega} \tag{306}$$

or

$$I(\omega) = 5\frac{1}{(2 + i\omega)(1 + i\omega)} \tag{307}$$

To obtain the inverse transform, we write the fraction on the right as

$$\frac{1}{(2 + i\omega)(1 + i\omega)} = \frac{1}{1 + i\omega} - \frac{1}{2 + i\omega} \tag{308}$$

Equation (305) shows that

$$FT[Ae^{-at}] = \frac{A}{a + i\omega} \tag{309}$$

Using this in the reverse direction, we have

$$i(t) = FT^{-1}[I(\omega)] = 5FT^{-1}\left[\frac{1}{1 + i\omega} - \frac{1}{2 + i\omega}\right] \tag{310}$$

and the solution is

$$i(t) = \begin{bmatrix} 0 & for\ t<0 \\ 5(e^{-t} - e^{-2t})\ for\ t\geq 0 \end{bmatrix} \tag{311}$$

as we previously obtained. It tells us that, at $t = 0$, the current starts to rise; it reaches a maximum and then decays. This applied voltage is called the *excitation*, and the resulting current is the *response*. These two quantities are called *time domain* variables. Their *frequency domain* forms $V(\omega)$ and $I(\omega)$ are given by (305) and (307), respectively, which combine as

$$I(\omega) = V(\omega)\left[\frac{1}{2(2 + i\omega)}\right] \tag{312}$$

It is known from circuit theory that the impedance of a resistor and an inductor in series is

$$Z(\omega) = R + i\omega L$$

which is a frequency domain expression. The second term on the right comes from the relation

$$v = L\, di(t)/dt$$

connecting voltage and current in an inductor. Writing the current as

$$i = i_{max}e^{i\omega t}$$

the previous equation becomes

$$v = Li\omega$$

by direct substitution. For this circuit, the *admittance* $Y(\omega)$ is

$$Y(\omega) = 1/Z(\omega) = 1/(4 + i\omega 2)$$

which we recognize as the term in square brackets in (312), so that

$$I(\omega) = Y(\omega)V(\omega) \tag{313}$$

and this is Ohm's law. The admittance is also known as the *transfer function*. It is a polynomial that acts on the excitation to generate the response. It also indicates how the impedance or admittance of the circuit depends on frequency. For example, note that, for $\omega = 0$, $Y(0) = 1/4$ ohm; that is, the circuit is simply a 4-ohm resistor, as expected. At high frequencies, the imaginary term is very large, and $Y(\omega)$ falls off as ω increases. We shall now show that (313), when expressed in the time domain, becomes

$$i(t) = \int_{-\infty}^{\infty} y(\alpha)v(t - \alpha)\,d\alpha \tag{314}$$

where α is an undefined variable with the dimensions of time. This result is known as the *convolution theorem* and it may be written as

$$i(t) = y(t) \circledast v(t) \tag{315}$$

where the special symbol denotes the *convolution* of $y(t)$ and $v(t)$. Taking the Fourier transform of $i(t)$ in (314) and reversing the order of integration, we obtain

$$I(\omega) = \int_{-\infty}^{\infty}\left[\int_{-\infty}^{\infty} y(\alpha)v(t - \alpha)d\alpha\right]e^{-i\omega t}dt \tag{316}$$

or

$$I(\omega) = \int_{-\infty}^{\infty}\left[\int_{-\infty}^{\infty} v(t - \alpha)e^{-i\omega t}dt\right]y(\alpha)d\alpha \tag{317}$$

Now let $\beta = t - \alpha$, where β is another variable like α. Holding α constant during the integration gives

$$I(\omega) = \int_{-\infty}^{\infty}\left[\int_{-\infty}^{\infty} v(\beta)e^{-i\omega\beta}d\beta\right]e^{-i\omega\alpha}y(\alpha)d\alpha \tag{318}$$

or

$$I(\omega) = FT[v(\beta)]\int_{-\infty}^{\infty} y(\alpha)e^{-i\omega\alpha}d\alpha = Y(\omega)V(\omega) \tag{319}$$

which agrees with (313). We have thus shown that the transfer function operating on the excitation in the frequency domain gives the response. The corresponding connection between these three quantities in the time domain is through a convolution. These concepts will be used to determine how the wavelength of the light passing through an optical system and the periodic properties of the object determine the contrast and resolution of the image.

CHAPTER 9

Fourier Optics

9.1 The Optical Transfer Function

I explained the significance of the transfer function of a circuit in Chapter 8. It tells us how the circuit will respond to an applied excitation. The same idea can be carried over to optics, and we shall see how the optical equivalent of a transfer function takes the excitation—the object—and produces a response—the image. We start this involved calculation with the asymmetric lens of Figure 9.1. Suppose that the lens is diffraction limited (although such a lens does not exist). Geometrical optics predicts that it will produce an image in accordance with paraxial optics, whereas wave optics shows that the circular rim of this lens produces an Airy disk. We combine geometrical optics and diffraction theory to determine the ability of such a lens to resolve two closely spaced points. Light leaves an object, passes through this lens, and forms an image. We find the amplitude at the lens by using the Huygens integral of equation (229) with the object as the source of light, and then we determine the amplitude at the image by integrating a second time over the lens area. The most complicated step is determining what happens at the lens. A ray of length T_1 leaves the object and strikes surface 1. The horizontal distance of the intersection point from the vertex, as measured along the axis, is

$$\overline{V_1 A_1} = r_1 \left[1 - \sqrt{1 - \frac{x_1^2 + y_1^2}{r_1^2}} \right] \tag{320}$$

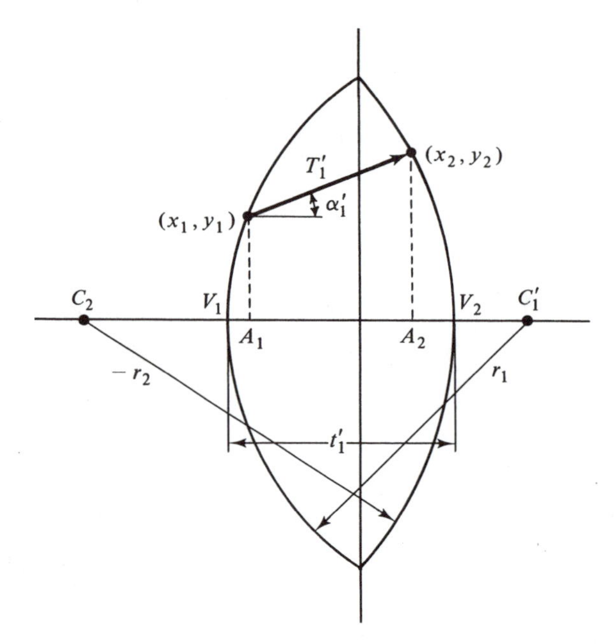

Fig. 9.1

Expanding the radical with a binomial approximation in the form $(1 + x)^{1/2} = 1 + {}^1\!/_2 x$, we obtain

$$\overline{V_1 A_1} = \frac{x_1^2 + y_1^2}{2r_1} \tag{321}$$

A similar result is obtained for the distance from A_2 to V_2, remembering to reverse the sign of r_2. If the slope of the refracted ray is small, then its length T_1' is obtained by subtracting the two quantities just calculated from t_1'. Yet another approximation is to replace both x_1 and x_2 with x_L, and similarly y_1 and y_2 with y_L, where x_L, y_L are the coordinates of any point on the ray inside the lens. These steps give

$$T'_1 = t'_1 - \left[\frac{1}{r_1} - \frac{1}{r_2}\right]\frac{x_L^2 + y_L^2}{2} = t'_1 - \frac{x_L^2 + y_L^2}{2f'(n'_1 - 1)} \tag{322}$$

where the last expression involves the use of the paraxial formula for the focal length. Returning to Figure 7.5, it will be seen that the term $x \sin\theta$ that appears in the diffraction integral represents a difference in the travel time of the two rays shown. This difference results in a phase shift, since the term ikr in (241) is called the phase of the complex quantity as written in polar form. We have a similar phase shift for the wave going through the lens. In air, it is the difference $(t'_1 - T'_1)$ between the axial thickness and the approximate thickness T'_1 for any off-axis ray. Inside the lens, the phase delay is due to the lower velocity in glass and is $n'_1 T'_1$. The total phase shift is the sum of these two quantities and, by substitution from (322), it becomes

$$n'_1T'_1 + (t'_1 - T'_1) = n'_1t'_1 - \frac{x_L^2 + y_L^2}{2f'} \tag{323}$$

The term $n'_1t'_1$ is a constant and can be removed from the integral; only the second term on the right need be kept. To it, we add the slant distance T'_2 from surface 2 to the image point (x',y'), which originally will be

$$T'_2 = \sqrt{(x' - x_L)^2 + (y - y_L)^2 + t'^2_2} \tag{324}$$

Again using the binomial theorem, this becomes

$$T'_2 = t'_2\left[1 + \frac{(x' - x_L)^2}{2t'^2_2} + \frac{(y' - y_L)^2}{2t'^2_2}\right] \tag{325}$$

and we have found the two components that go into the exponent of the Huygens integral at the lens. There will also be a corresponding expression for the object-to-lens distance T_1, but it will have a negative sign because T_1 and t_1 are opposite in sign. We integrate over the plane of the object using T_1 in the exponent of the diffraction integral to obtain the amplitude of the wave received at the lens, and then we apply Huygens' principle a second time. The resulting integration over both the lens area and the object area produces the expression

$$A(x',y') = \\ C\iint A(x,y)e^{-ik\frac{(x - x_L)^2 + (y - y_L)^2}{2t_1}}e^{-ik\frac{(x_L^2 + y_L^2)}{2f'}+ik\frac{(x' - x_L)^2 + (y' - y_L)^2}{2t'_2}}dS_LdS \tag{326}$$

where the constant terms involving t_1, t'_1, and t'_2 have been taken outside of the integral. Expanding the terms $(x - x_L)^2$ and $(y - y_L)^2$, the quantities x^2, y^2, x'^2 , and y'^2 are constant for a spherical object producing a spherical image, and nearly so for plane objects. When the three terms in x_L^2 are combined, the exponent will be zero if equation (76), the Gauss lens equation, is obeyed; a similar statement is true for the y_L^2 terms. The integral that is left is

$$A(x',y') = C\iint A(x,y)e^{ik\left[x_L\left(\frac{x}{t_1} - \frac{x'}{t'_2}\right) + y_L\left(\frac{y}{t_1} - \frac{y'}{t'_2}\right)\right]}dS_LdS \tag{327}$$

It is now necessary to introduce a basic property of the object, illustrated in Figure 9.2. This is a bar chart that is used to determine the resolution obtainable with an optical system. A series of uniformly spaced dark lines is specified in terms of the *spatial frequency*, defined as the number of line-pairs or cycles per unit length, and expressed in units of line-pairs/m or—more commonly—line-pairs/mm. *Temporal frequency* v, which describes a propagating wave, is the number of cycles per second (or hertz), whereas spatial frequency v_x applies to a stationary object. In analogy with the

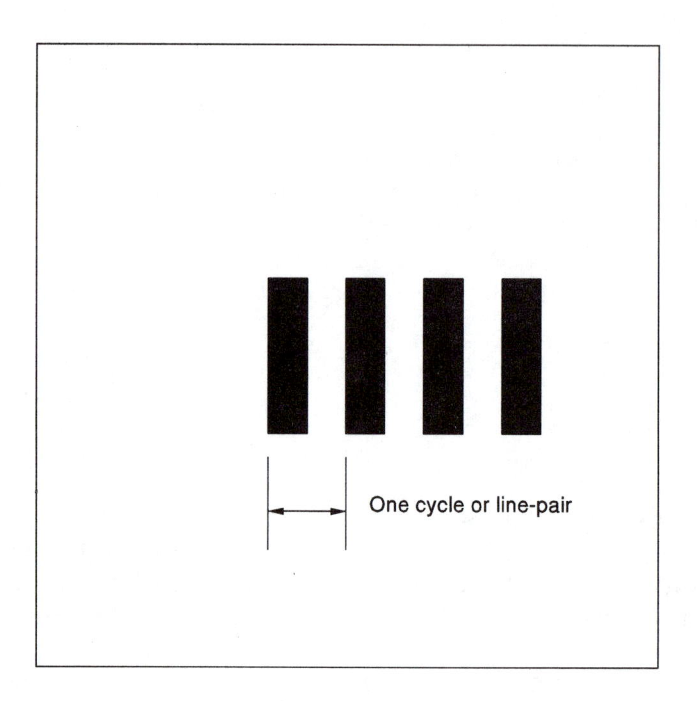

Fig. 9.2

relation between frequency and angular frequency, a *spatial angular frequency* ω_x would be related to the spatial frequency through the expression

$$\omega_x = 2\pi\nu_x$$

To convert (327) into an integral involving the response of the lens to a range of spatial frequencies, we substitute the variables

$$\omega_x = \frac{kx_L}{t_2'} \qquad \omega_y = \frac{ky_L}{t_2'} \tag{328}$$

We also use the paraxial relation $t'_2 = mt_1$, where m is the magnification, giving

$$A(x',y') = C\iiint A(x,y)e^{i[\omega_x(mx - x') + \omega_y(my - y')]}d\omega_x d\omega_y dS \tag{329}$$

The use of the Gaussian equation implies a perfect lens. For a lens with aberrations, we introduce a *pupil* or *aperture function* $P(x_L, y_L)$, or $P(\omega_x, \omega_y)$. This function was derived in Chapter 5; it is of the form exp (ikw), where w is a measure of the phase shift due to the aberrations. We need only realize that w will be zero for a diffraction-limited system, so that the pupil function has a value of unity inside the lens and zero outside. Then the integral is

$$A(x',y') = C\iiint P\left(\frac{\omega_x t'_2}{k},\frac{\omega_y t'_2}{k}\right)A(x,y)e^{i[\omega_x(mx - x') + \omega_y(my - y')]}d\omega_x d\omega_y dS \tag{330}$$

This can be simplified if we define a function

$$h(mx - x', my - y') = C\int\int P\left(\frac{\omega_x t'}{k}_2, \frac{\omega_y t'}{k}_2\right) e^{i[\omega_x(mx - x') + \omega_y(my - y')]} d\omega_x d\omega_y \tag{331}$$

so that the diffraction integral finally becomes

$$A(x', y') = \int A(x, y) h(mx - x', my - y') dS \tag{332}$$

This result looks something like a convolution, but there are two sets of variables, and the term $h(mx - x', my - y')$ does not involve a reflection in the vertical axis. This is of no importance for the applications we shall consider shortly; they involve symmetrical functions and reflection has no effect on the outcome. Hence, we accept (332) as a convolution, with $h(mx - x', my - y')$ as the optical equivalent of the admittance $y(t)$; it determines the image response $A(x', y')$ caused by the excitation $A(x, y)$. Since most systems are not diffraction limited, the image will be an Airy disk that is distorted in some way, so that $h(mx - x', my - y')$ is called the *point-spread function*; its use in (332) gives the response to the excitation at the point (x, y). Letting $x = y = 0$ (for a point object on the axis) and reversing the sign on x' and y', (331) becomes

$$h(x', y') = C\int\int P\left(\frac{\omega_x t'}{k}_2, \frac{\omega_y t'}{k}_2\right) e^{i(\omega_x x' + \omega_y y')} d\omega_x d\omega_y \tag{333}$$

showing that $h(x', y')$ is the inverse Fourier transform of the pupil function. Because h is the response to a point source, it is sometimes called the *impulse response function*, and its Fourier transform $H(\omega_x, \omega_y)$ is the *optical amplitude transfer function*. Before proceeding further, however, let's consider the meaning of the convolution process.

9.2 CONVOLUTION AS AN AREA OVERLAP PROCEDURE

Consider the set of two rectangular pulses in Figure 9.3a as well as the set of three pulses in Figure 9.3b; the amplitudes of these pulses are functions $g(\alpha)$ and $h(\alpha)$, respectively, of an arbitrary variable α. By (314), their convolution is

$$f(t) = \int g(\alpha) h(t - \alpha) d\alpha \tag{334}$$

This equation tells us that the convolution is obtained by reversing the function $h(\alpha)$ (Figure 9.3c) and then shifting $h(-\alpha)$ by an amount t, chosen in this case as $t = -2$, as shown in Figure 9.3d. The corresponding values of $g(\alpha)$ and $h(t - \alpha)$ are multiplied to obtain the product shown in Figure 9.3e. The total shaded area amounts to 2.0 units; this is plotted in Figure 9.3f.

PROBLEM 104

Complete the explanation of Figure 9.3f.

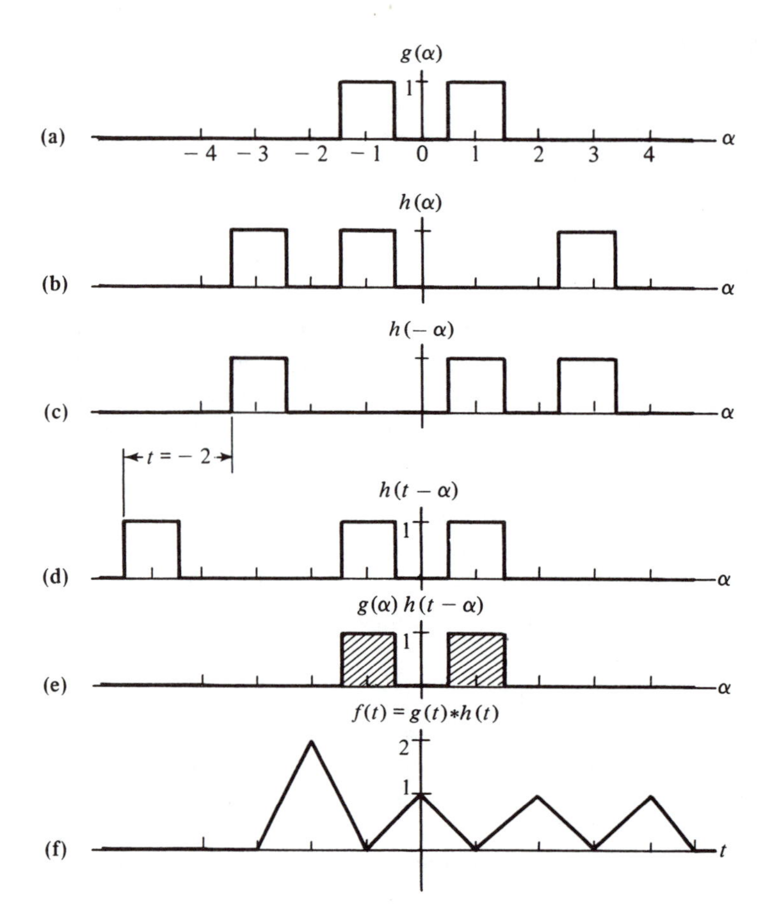

Fig. 9.3

This convolution has been generated optically with an ingenious piece of apparatus designed by R. E. Haskell (see *IEEE Transactions on Education* E-14, 110 [1971]). A laser beam (Figure 9.4a) is expanded and forms an image of a three-slit pattern at the plane where the two-slit pattern $g(\alpha)$ is located. Lens 1 gives an inverted image, which is equivalent to changing $h(\alpha)$ to $h(-\alpha)$. The vibrating mirror sweeps $h(-\alpha)$ across $g(\alpha)$, thus performing the convolution, and the received signal is displayed on an oscilloscope swept by the same signal that drives the speaker. The resulting trace (Figure 9.4b) agrees very well with Figure 9.3f.

9.3 The Optical Intensity Transfer Function

The optical amplitude transfer function $H(\omega_x,\omega_y)$ was obtained by combining diffraction theory and geometrical optics. What we really want is the *optical intensity transfer function* $F(\omega_x,\omega_y)$, known simply as the *optical transfer function (OTF)*. An optical system takes the energy from a source of light and forms an image. Since the energy

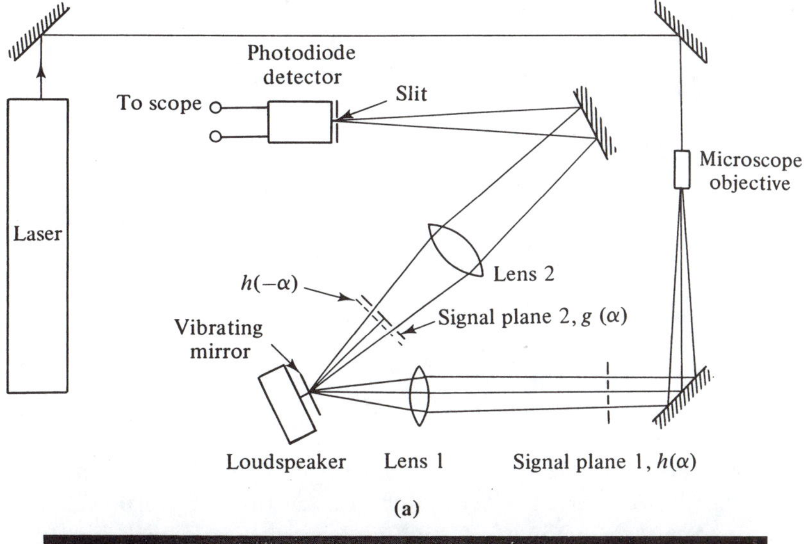

(a)

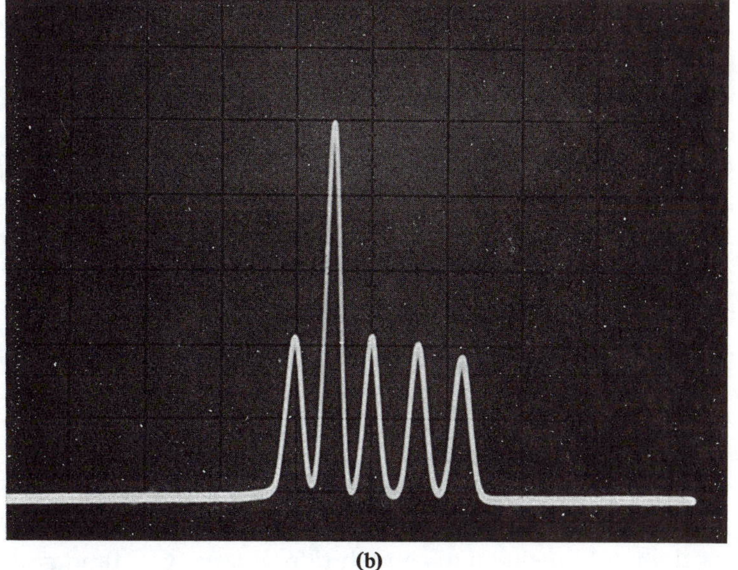

Fig. 9.4

(b)

of an electromagnetic wave depends on the square of the amplitude A, or on the product AA^* for a complex quantity, we define the OTF as

$$F(\omega_x,\omega_y) = FT[hh^*] \tag{335}$$

where h^* is the complex conjugate of h. This Fourier transform involving two variables is then

$$F(\omega_x,\omega_y) = \iint h(\xi,\eta)h^*(\xi,\eta)e^{-i(\omega_x\xi + \omega_y\eta)}d\xi d\eta \tag{336}$$

where ξ and η are arbitrary coordinates. Since h^* is the inverse transform of H^*, then

$$F(\omega_x,\omega_y) = \iint h(\xi,\eta)\left[\iint H^*(\omega'_x,\omega'_y)e^{i(\xi\omega'_x+\eta\omega'_y)}d\omega'_x d\omega'_y\right]e^{-i(\omega_x\xi+\omega_y\eta)}d\xi d\eta \qquad (337)$$

where ω'_x and ω'_y are another arbitrary pair of variables and the constant factor of 1/ 2π that goes with the inverse transform is ignored. Rearranging gives

$$F(\omega_x,\omega_y) = \iiiint H^*(\omega'_x,\omega'_y)h(\xi,\eta)e^{-i[(\xi(\omega_x - \omega'_x) + \eta(\omega_y - \omega'_y)]}d\omega'_x d\omega'_y d\xi d\eta \qquad (338)$$

or

$$F(\omega_x,\omega_y) = \iint H^*(\omega'_x,\omega'_y)H(\omega_x - \omega'_x,\omega_y - \omega'_y)d\omega'_x d\omega'_y \qquad (339)$$

Since the point-spread function h is the inverse transform of the transfer function H and, by (333), it is also the inverse transform of the pupil function P, then H and P must be the same except for a constant, and we can replace H with P to obtain

$$F(\omega_x,\omega_y) = \iint P^*\left(\frac{\omega'_x t'_2}{k},\frac{\omega'_y t'_2}{k}\right)P\left(\frac{(\omega_x-\omega'_x)t'_2}{k},\frac{(\omega_y-\omega'_y)t'_2}{k}\right)d\omega'_x d\omega'_y \qquad (340)$$

The signs on the differences $(\omega_x - \omega'_x)t'_2/k$ and $(\omega_y - \omega'_y)t'_2/k$ can be reversed without affecting the absolute value of the integral, and ω'_x and ω'_y can be replaced with x_L and x_Y, respectively, using the definitions of (328). Then the shift in origin given by

$$x_L = x'_L - \frac{\omega_x t'_2}{2k},\quad y_L = y'_L - \frac{\omega_y t'_2}{2k} \qquad (341)$$

produces the expression

$$F(\omega_x,\omega_y) = \iint P\left(x_L + \frac{\omega_x t'_2}{2k}, y_L + \frac{\omega_y t'_2}{2k}\right)P^*\left(x_L - \frac{\omega_x t'_2}{2k}, y_L - \frac{\omega_y t'_2}{2k}\right)dx_L dy_L \qquad (342)$$

This equation is the two-dimensional convolution of two overlapping areas, one centered at $(\omega_x t_2'/2k, \omega_y t_2'/2k)$ and the other at the diametrically opposite point $(-\omega_x t_2'/2k, -\omega_y t_2'/2k)$. The areas are circles of radius r corresponding to the lens rim (Figure 9.5). Without loss of generality, we can place the centers on an axis and deal with a one-dimensional problem. Although the pupil function P is a constant within the rim of the lens, the integral must be evaluated like any other convolution; that is, by calculating how the overlap area changes as one circle passes over the other. This overlap area is the shaded portion of Figure 9.5a, and four times the area of the part labeled S in Figures 9.5a and b. The area A of the sector of angle θ and radius r is

$$A = (\theta/2\pi)\pi r^2 = \theta r^2/2$$

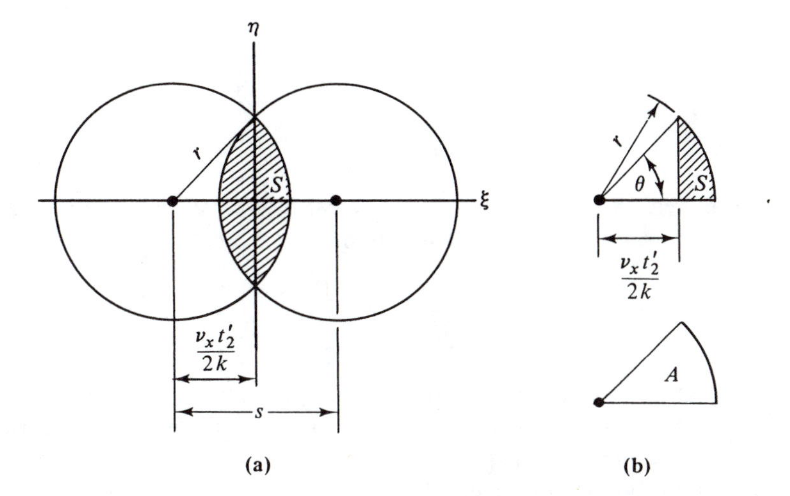

Fig. 9.5

where

$$\theta = \text{arc cos } (s/2r)$$

and the spacing is

$$s = \omega_x t'_2/k$$

The two parts of Figure 9.5b show that S has an area

$$S = A - \frac{1}{2}\left(\frac{s}{2}\right)\sqrt{r^2 - \left(\frac{s}{2}\right)^2} \tag{343}$$

Then, using the above expressions for A and θ, the convolution integral of (341) has a value

$$F(\omega_x,0) = \frac{4S}{\pi r^2} = \frac{2}{\pi}\left[\text{arc cos}\frac{s}{2r} - \frac{s}{2r}\sqrt{1 - \left(\frac{s}{2r}\right)^2}\right] \tag{344}$$

where $0 \le s/2r \le 1$. The total overlap area is divided by the area of the circle so that the maximum value of the transfer function, occurring at $s = 0$, will be unity. This expression indicates that $F(\omega_x,0)$ is defined only for $s \le 2r$ and that it vanishes for $s = 2r$, which is the condition for the circles just touching. The corresponding value ω_{x0} of ω_x is called the *cutoff angular frequency.* In our treatment of circuits at the end of Chapter 8, we saw that the transfer function $Y(\omega)$ depends on the angular frequency ω of the excitation in such a way as to have its maximum for $\omega = 0$ and to become very small when ω is large. We would therefore expect that the OTF will describe how the lens responds to the spatial frequency of the object and that the variable $s/2r$ is a measure of this frequency. To confirm this, we use the definition of s shown in Figure 9.5a to obtain

$$s/2r = \omega_x t'_2/2rk$$

Since the transfer function vanishes when $s/2r = 1$, then

$$\omega_{x0} = 2rk/t'_2$$

and

$$s/2r = \omega_x/\omega_{x0} = \nu_x/\nu_{x0}$$

so that $s/2r$ can be called a *relative spatial frequency*, and ν_{x0} represents the maximum spatial frequency that the lens can resolve. It is given by

$$\nu_{x0} = \frac{\omega_{x0}}{2\pi} = \frac{2rk}{2\pi t'_2} = \frac{2r}{\lambda t'_2} = \frac{1}{\lambda f\#} \tag{345}$$

where, for an object at infinity, the ratio of image position to diameter is the f-number. A plot of the transfer function as it varies with spatial frequency is shown in Figure 9.6a (solid curve). The dotted curve is the transfer function for a rectangular opening, which is linear since the overlap area of one rectangle passing over another goes from zero to a maximum. Figure 9.6b is the usual way of showing the transfer function. To estimate the resolution obtainable with a diffraction-limited lens, we use equation (345). For example, an f:2.8 lens used with green light has a cutoff spatial frequency $\nu_{x0} = 1/[0.55 \times 10^{-3}\,(\text{mm}) \times 2.8] = 650$ line-pairs/mm, so that a single line-pair has a width of somewhat less than 2 μm. This value shows that a diffraction-limited lens would be too good for a 35 mm camera, since most commercial film has a resolution of about 10 μm. W. E. Smith in *Modern Optical Engineering*, McGraw-Hill (1990), states that excellent readability requires the ability to resolve 8 line-pairs per width of lowercase "e," and 3 line-pairs would be barely distinguishable.

PROBLEM 105

Find the transfer function for a square aperture of side $d = 2r$, where r is the radius of the circular aperture of Figure 9.5. Verify the upper curve in Figure 9.6a and the fact that ν_{x0} is the same in both cases.

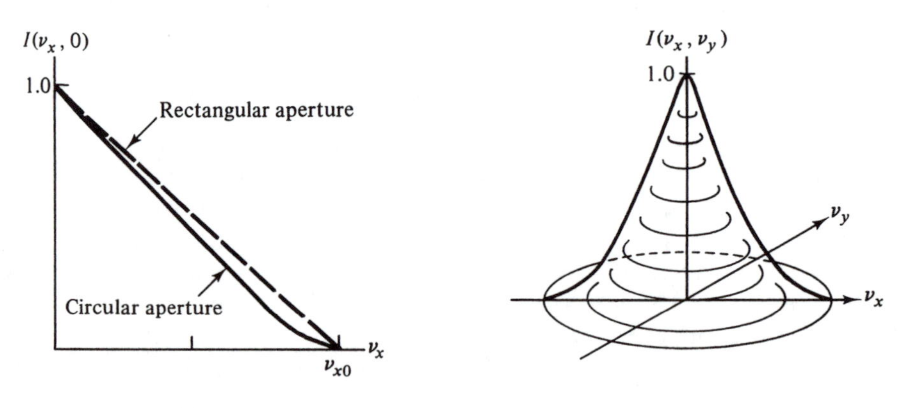

Fig. 9.6

A simplification of the above results is obtained by letting $s/2r$ in (344) be small. The power series for the inverse cosine is

$$\text{arc cos x} = \pi/2 - \text{x} - \ldots$$

and the second term reduces to $s/2r$ so that the transfer function becomes

$$\text{F} = 1 - 4\nu_r/\pi$$

where $\nu_r = s/2r$ is the relative spatial frequency. This indicates that the part of the transfer function near the origin in Figure 9.6a can be approximated by the straight line shown in Figure 9.6b. It strikes the horizontal axis at a value $\nu_r = \pi/4$ or 0.785, corresponding to F = 0, and the cutoff is thus estimated as being about 20% below its true value.

9.4 THE MODULATION TRANSFER FUNCTION (MTF)

The theoretical treatment of the transfer function just given applies only to a diffraction-limited system. When aberrations are present to a moderate or strong degree, the corresponding calculation is very complicated, and it is much easier to use a procedure that is based in part on ray tracing, as will now be developed. The bar chart of Figure 9.2 can be regarded as a periodic array of rectangular pulses that can be described by a Fourier series. Let each component of the series have an intensity of the form

$$I(x) = b_0 + b_1 \cos k_x x \tag{346}$$

as illustrated in Figure 9.7. Note that b_0 represents the average value of this component and that the maximum and minimum values are $(b_0 + b_1)$ and $(b_0 - b_1)$, respectively. The variables λ_x and k_x used in this section are *spatial* quantities and are unrelated to the wavelength of the light, which comes in later. Define the *contrast* or *modulation* M of this component as

$$M = \frac{I_{max} - I_{min}}{I_{max} + I_{min}} \tag{347}$$

Substituting the maximum and minimum values into (346), the modulation of the object is

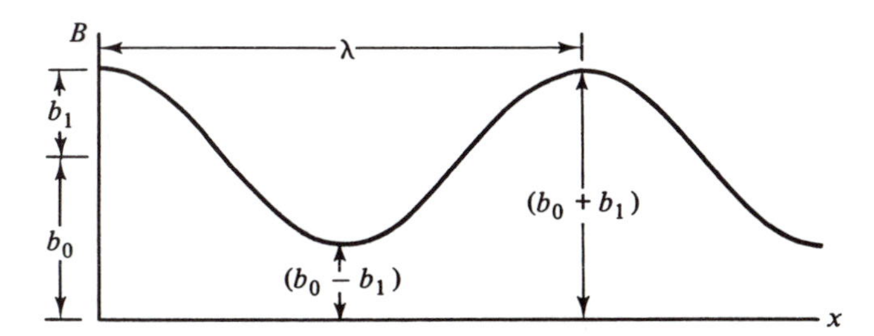

Fig. 9.7

$$M_{ob} = \frac{b_1}{b_0} \tag{348}$$

To determine the modulation of the image, simplify the procedure by considering a one-dimensional problem. Equation (332) indicates that the image amplitude is the convolution in *coordinate space* of the object amplitude and the point-spread function $h(x', y')$. To do the same thing for intensities, we use the inverse Fourier transform $f(x', y')$ of the OTF $F(\nu_x, \nu_y)$ and write the convolution as

$$I(x') = \int f(x)[b_0 + b_1 \cos k_x(x' - x)]\, dx \tag{349}$$

where $f(x)$ is now called a *line-spread function*. Expanding this, we obtain

$$I(x') = b_0 \int f(x)\, dx + b_1 \int f(x) \cos k_x x' \cos k_x x\, dx + b_1 \int f(x) \sin k_x x' \sin k_x x\, dx \tag{350}$$

Dividing the equation by the integral in the first term on the right, and letting $I(x')$ denote the new intensity after this normalization, (350) becomes

$$I(x') = b_0 + b_1 \cos k_x x' \frac{\int f(x) \cos k_x x\, dx}{\int f(x) dx} + b_1 \sin k_x x' \frac{\int f(x) \sin k_x x\, dx}{\int f(x) dx} \tag{351}$$

or

$$I(x') = b_0 + b_1 F_R \cos k_x x' + b_1 F_I \sin k_x x' \tag{352}$$

where F_R and F_I are the real and imaginary parts of the complex quantity

$$F(-k_x) = \frac{\int f(x) e^{-ik_x x} dx}{\int f(x)\, dx} \tag{353}$$

By using the negative sign, we can identify the numerator as the Fourier transform of f; that is, F and f are related as they should be, aside from the scale-adjusting constant in the denominator. Then

$$I(x') = b_0 + b_1 |F| \cos(k_x x' - \phi) \tag{354}$$

where the magnitude of F is

$$|F| = \sqrt{F_R^2 + F_I^2} \tag{355}$$

and the phase angle is

$$\phi = \text{arc}\ \tan \frac{f_I}{f_R} \tag{356}$$

We are specifically interested in the magnitude of $f(k)$. By (354), the maximum and minimum values of the intensity at the image plane will occur when $\cos(kx' - \phi)$ is either 1 or −1, respectively, so that

$$I_{max}(x') = b_0 + b_1|F| \quad (357)$$

and

$$I_{min}(x') = b_0 - b_1|F| \quad (358)$$

Then (348) gives

$$M_{im} = \frac{b_1}{b_0}|F| \quad (359)$$

and

$$|F| = \frac{M_{im}}{M_{ob}} \quad (360)$$

The magnitude of F is called the *modulation transfer function*; it expresses as a ratio the amount by which the lens degrades the modulation or contrast of the bar target. If there is no degradation, this ratio is unity or 100%.

The MTF is found from information available in the spatial domain. This is the region for which the variable is x. A regular pattern formed by the intersection of equally spaced concentric circles and radial lines is used as an object. Rays are traced from each of these points to the paraxial focal plane, with Figure 9.8 showing the resulting image distribution; this *spot diagram* was produced by the triplet in Table 9.1.

The diagram in Figure 9.8 shows how the triplet will act on a uniform object to create an image with aberrations. Hence, we label this distribution $f(x)$ and recognize it as the transfer function as it exists in the spatial domain. The spot diagram was

Table 9.1 Triplet Specifications

r	*t′*	*n′*
1.63255		
	0.357	1.518522
−22.97151		
	0.189	1.0
−1.27303		
	0.081	1.629598
−3.20300		
	0.325	1.0
1.88311		
	0.396	1.629598
−2.45567		

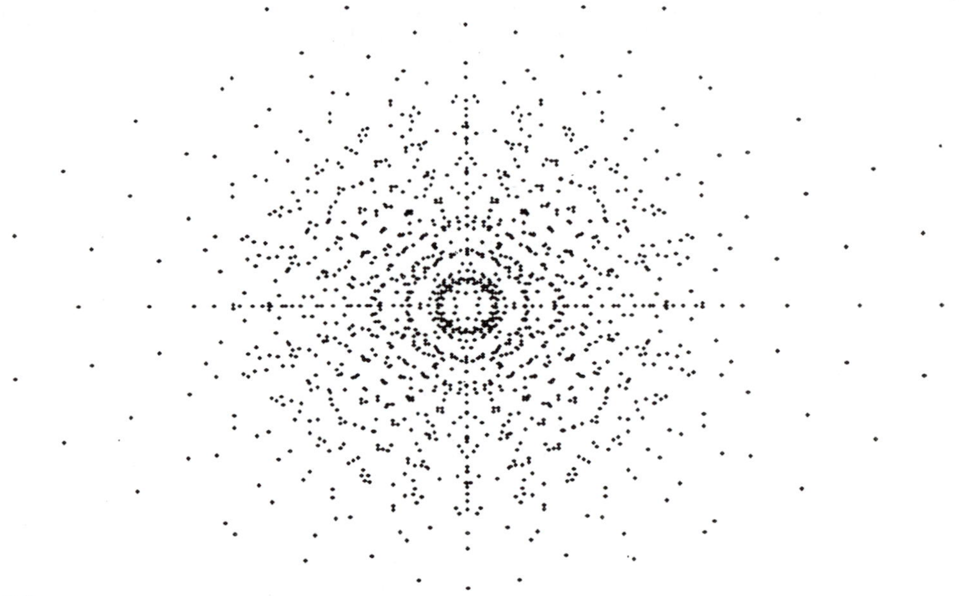

Fig. 9.8

generated by placing the uniform object at 15 units from the triplet. The object points were formed by the intersection of 8 circles with radii 0.02, 0.07, 0.12, 0.17, 0.22, 0.27, 0.32, 0.37 and 16 radial lines for a total of 128 points. A vertical fan of 5 rays and a horizontal fan of 5 rays left each point, so that 1,280 rays were traced. The ray tracing program incorporated a procedure to divide the diagram into equal slices parallel to the y-axis, and then the number of spots per slice was counted. Although $F(k_x)$ is normally computed as the Fourier transform of $f(x)$ by integration, in this situation we do not have an analytic expression for $f(x)$. Instead, we use the *discrete Fourier transform*, which approximates the integrals as sums. To apply the discrete transform, we need a simple procedure for dividing the spot diagram into slices and counting the number of image points in each slice. The diagram is symmetric about the y-axis, so that we need count only the spots lying in the right-hand half. A comparison procedure in the program indicated that the spot with the largest x coordinate was at x = 0.13. Multiplying this number by 200 converts the coordinate range into a series of integers ranging from 0 to 26. The location of any point in a series of slices that are 1 unit wide can then be determined by using the BASIC operation INT(200*X), which returns the greatest integer less than or equal to the quantity in parentheses. The resulting distribution of points is shown in Table 9.2.

The cosine term in the numerator of (353) will be

$$\int f(x)\cos kx\,dx = \Sigma f(x)\cos(kx)\,\Delta x \tag{361}$$

and the integral involving $\sin(k_x x)$, an odd function, vanishes. The argument is rewritten as

Table 9.2

Points per Slice	Slice Number
157	1
89	2
70	3
67	4
53	5
51	6
38	7
34	8
27	9
20	10
8	11
13	12
1	13
2	14
4	15
2	16
0	17
1	18
0	19
0	20
0	21
2	22
0	23
0	24
0	25
Total 639	

$$\cos(k_x x) = \cos(2\pi \nu_x x)$$

The denominator is

$$\int f(x)dx = \Sigma f(x)\Delta x \tag{362}$$

Let's calculate the sum of the cosine terms for a spatial frequency of 30 line-pairs/mm. They will have the form $\cos[2\pi(30)(K/200)] = \cos(0.06\pi K)$, where K = 1 to 22 (the terms beyond this do not contribute). The values of these terms are 0.982, 0.930, 0.844, 0.729, 0.588, 0.426, 0.249, 0.063, –0.125, –0.309, –0.482, –0.637, –0.771, –0.876, –0.951, –0.992, –0.998, –0.969, –0.905, –0.809, and –0.685. Each of these terms is multiplied by the corresponding number of spots per slice as given in Table 9.2, and a sum of 378.51 is obtained. Dividing by the total number of spots gives the MTF as 378.5/649 =

0.58 or 58%. A plot of the MTF for this triplet as a function of the spatial frequency is shown in Figure 9.9. As an example of a design problem to which these procedures have been applied, it was necessary to produce an optical system for medical purposes that had a fairly long focal length and a resolution of 10 μm. None of the lenses listed in catalogues met this requirement. After a search, the specifications for a camera lens called the Heliar were found in *Lens Design* by M. Laikin (Marcel Dekker Inc., 1991). This lens is like the Tessar described earlier, but with the single front element replaced by a cemented doublet. It was found that even this lens did not quite have the necessary resolution, but two of them placed back-to-back and nearly touching gave the three curves of Figure 9.10 for an off-axis object. Note that the system is almost diffraction limited for short and medium wavelengths; fortunately, this is the region for which photographic film is most sensitive. The limit of 100 line-pairs/mm on this graph corresponds to the desired resolution of 10 μm, which appears to occur at an MTF of about 50%. However, the system has a magnification of about 1.7, so that the actual limit comes at an MTF of 85%, a useful design criterion.

PROBLEM 106

Write a program that will reproduce Figure 9.9. Use the suggestions given in the preceding text and generate the axes and markings with the LINE command.

The transfer function discussed in this section is also known as the *geometrical optics transfer function* because of the way it is obtained. This distinguishes it from

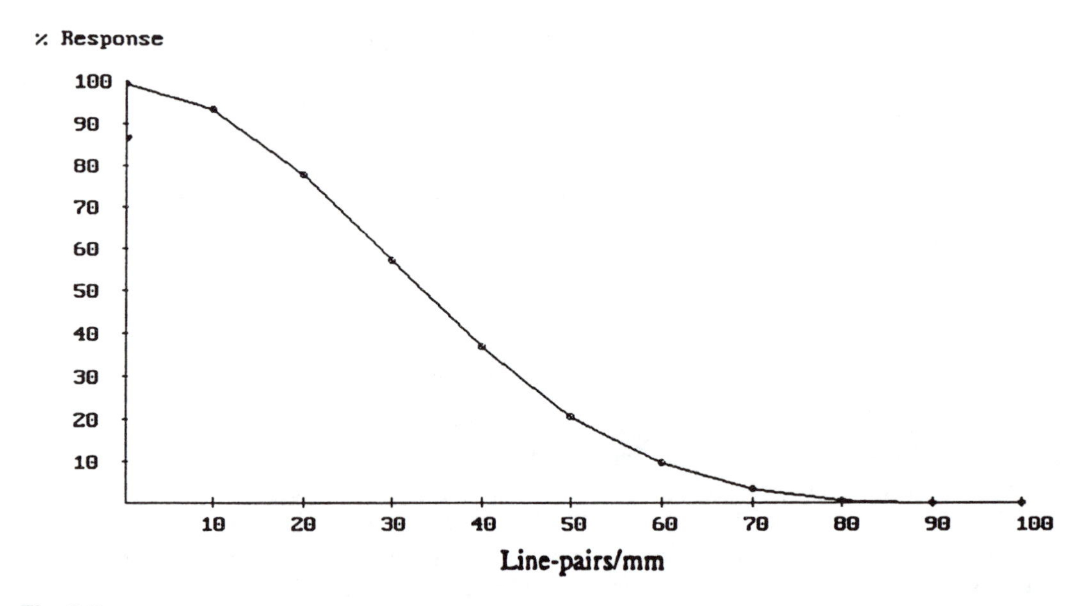

Fig. 9.9

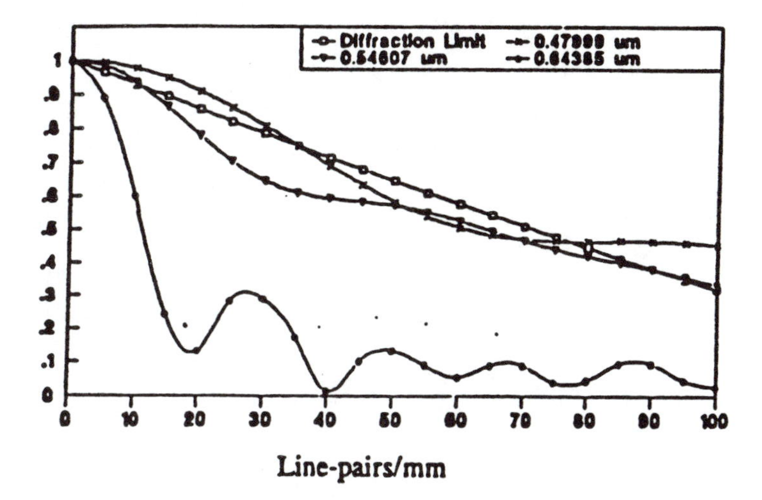

Fig. 9.10

the *diffraction-based transfer function*, which is calculated by an integration. Returning to the discussion of the transfer function for a circuit presented in Chapter 8, we recall that the effect of operating on the excitation with the transfer function in the frequency domain produces the response and that the corresponding connection between these three quantities in the time domain is through a convolution. This forms the basis for equation (349), with the excitation being the object, the response being the image, and the transfer function in the spatial domain being obtained from the distribution of points in the spot diagram. We have given here a simple procedure for showing how aberrations limit the resolution and contrast obtainable by an optical system.

CHAPTER 10

Light as a Form of Energy

10.1 ILLUMINATION AND RADIATION

The efficiency of an optical system in transferring energy is an important part of its design. Illumination or radiation problems are confusing because there are many similar quantities used to specify the amount of energy transferred or received as well as an enormous number of units for these quantities. In recent years, the situation has improved with the gradual changeover to the International System of Units (Système Internationale or SI units), which uses seven basic units of measurement: the meter, kilogram, and second are the fundamental units for dynamics; the ampere extends the coverage to electromagnetism and the kelvin to thermodynamics. The basic unit for quantity of matter is the mole (the molecular weight expressed in kilograms) and the seventh basic unit is the candela, which is used for light as a form of energy and which will be defined shortly. Then there are two auxiliary units involving angular measure, which we now review.

A *circular radian* or simply *radian* is the angle subtended at the center of a circle by an arc on its circumference of length equal to the radius r. Dividing $2\pi r$ by r gives 2π as the number of radians in a full circle. The number of degrees per radian is $360°/2\pi$ or $57.3°$. Extending these definitions to a sphere, a *spherical radian* or *steradian* is the *solid angle* subtended at its center by an area of magnitude r^2. Since the surface is $4\pi r^2$, there are 4π steradians surrounding a point in space. Let a cone of arbitrary shape have its apex at the center of a sphere of unit radius. The solid angle ω at the apex is numerically equal to the surface area on the sphere intercepted by the cone, since the full sphere has an area of 4π. Now consider a cone with its apex at the center of an arbitrary sphere. Let it have a vertex solid angle $d\omega$ and a base of area dS, with some point P near the center of the base lying on the surface of the sphere. Let the normal to the base make an angle θ with the radius passing through P. Then the projection of dS onto the spherical surface at P has an area $dS \cos \theta$. Since the solid angle is the area the cone intercepts on a unit sphere, then $dS \cos \theta/r^2 = d\omega/1^2$ or

$$d\omega = \frac{dS \cos \theta}{r^2} \tag{363}$$

PROBLEM 107

Consider a point P on a line normal to the center of a disk. The point is a distance a from the center of the disk, which has a radius R.

(a) Use an integration to show that the solid angle subtended by the disk at the point is

$$\omega = 2\pi a\left[\frac{1}{a} - \frac{1}{\sqrt{a^2 + R^2}}\right] \tag{364}$$

(b) Show that, when the point P is very far from the disk, the solid angle reduces to zero.

(c) Show that, when P is very close to the disk, the solid angle becomes 2π.

(d) Show that, when $R/a < 1$, (364) simplifies to

$$\omega = \pi R^2/a^2$$

(e) Explain physically what (b), (c), and (d) mean.

Turning now to light energy, there are five basic quantities that have to be introduced. Four of them concern the source and the fifth one deals with the receiver. Let the area dS in Figure 10.1 emit electromagnetic radiation toward a receiving area dS' at a distance r. The energy being transmitted can have any wavelength in the entire spectrum or it can be limited to the visible portion. If the former, we call it *radiant* energy and the units are *joules*. If it is in the 400–700 nm range, then it is *luminous*

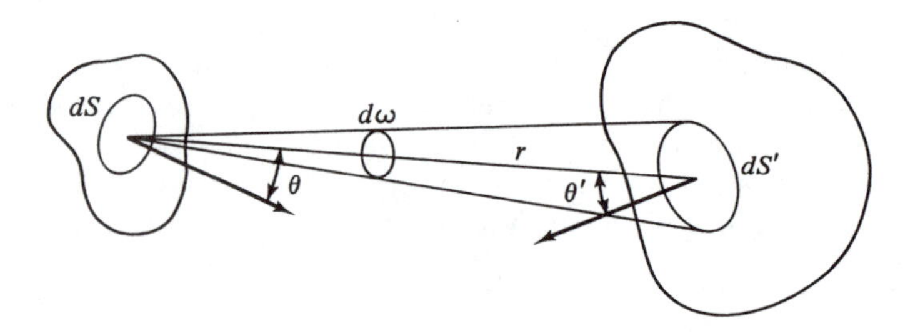

Fig. 10.1

energy, and the unit is the *talbot*. These quantities are at the top of Table 10.1. This table will summarize and help you understand our discussion. The table has two columns, and every concept appears twice. The left-hand column lists *radiometric* quantities, which apply to radiation of any wavelength, and the right-hand side list the corresponding *photometric* quantities for visible light.

The four sets of units that specify a quantity of radiation are

1. Power
2. Power per unit area
3. Power per unit solid angle
4. Power per unit solid angle per unit projected area

and the energy received is expressed as power per unit area. To be specific, the radiometric power dP emitted from a surface dS is called the *radiant flux* and is measured in watts. Then the power per unit area is the *radiant emittance* W, and it will be

$$W = dP/dS \text{ watts/meter}^2$$

To consider the corresponding photometric quantities, we have to bring in the properties of the human eye, since these are what determine the visibility of light. Figure 10.2 shows the sensitivity as a function of wavelength or color. The peak is in the green at 555 nm, and it was internationally agreed that 1 watt of green light is equivalent to 680 *lumens* (this was changed in 1980 to 683 lm).

The photometric equivalent of the radiant flux P is the *luminous flux* F, whose units are lumens, and the luminous flux per unit area is the *luminous emittance* L, measured in lumens/m^2.

Next we introduce the *radiant intensity* J, with a unit of watt/steradian and the *luminous intensity* I, for which the unit of the lumen/steradian is called the *candela* (once known as the *candle*), and I was called the "candlepower." The final pair of this first four is *radiance* N, measured in watt/steradian/m^2, and *luminance* B, in candela/m^2 (or lumen/steradian/m^2). Luminance B is the property of a source that is commonly

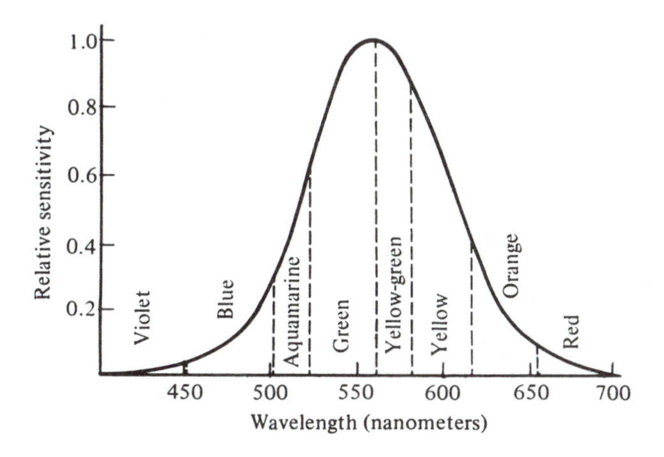

Fig. 10.2

Table 10.1 Summary of SI (and Other) Units of Radiant Energy

Radiometric	Photometric
Radiant energy U joule	Luminous energy Q talbot
Radiant flux P watt	Luminous flux F lumen
Flux is total output power ("light"). 1 W = 680 lm at 555 nm	
Radiant emittance W watt meter^{-2}	Luminous emittance L lumen meter^{-2}
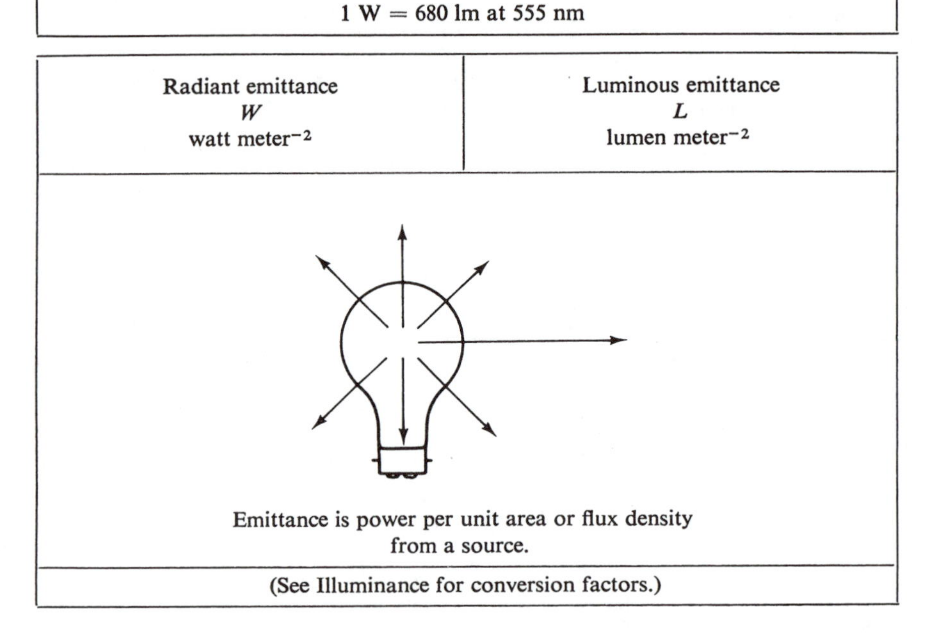 Emittance is power per unit area or flux density from a source.	
(See Illuminance for conversion factors.)	

Table 10.1 Continued

<table>
<tr><td>Radiant intensity
J
watt steradian^{-1}</td><td>Luminous intensity
I
candela
(lumen steradian^{-1}, candle)</td></tr>
<tr><td colspan="2">Intensity is power per unit solid angle ("candlepower").
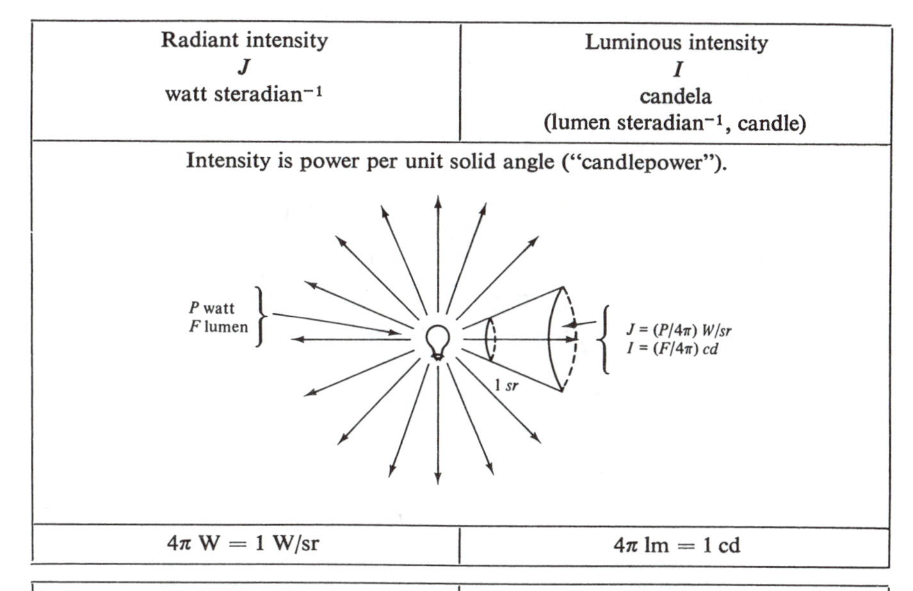
</td></tr>
<tr><td>4π W = 1 W/sr</td><td>4π lm = 1 cd</td></tr>
</table>

<table>
<tr><td>Radiance
N
watt steradian^{-1} meter^{-1}</td><td>Luminance
B
candela meter^{-2}
(lumen steradian^{-1} meter^{-2},
lambert)</td></tr>
<tr><td colspan="2">Radiance or luminance is power per unit solid angle per unit projected area emitted or scattered ("brightness").
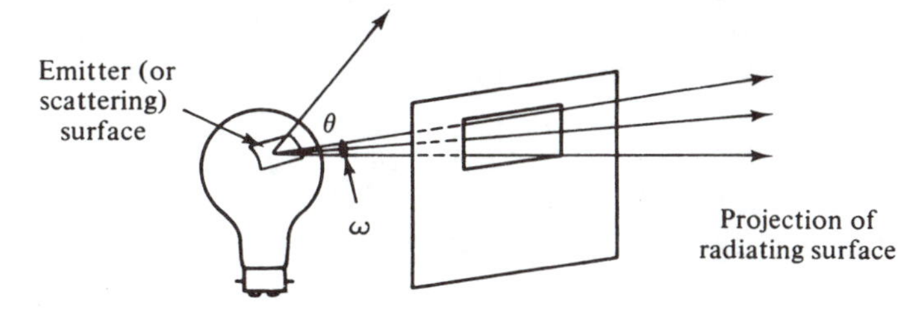
</td></tr>
<tr><td colspan="2">For a Lambertian source only:
B cd/m^2 = πL lm/m^2
1 L = 1 lm/cm^2
1 m-L = 1 lm/m^2
Also:
1 m-L = (0.0929 m^2/ft^2) ft-L</td></tr>
</table>

Table 10.1 Continued

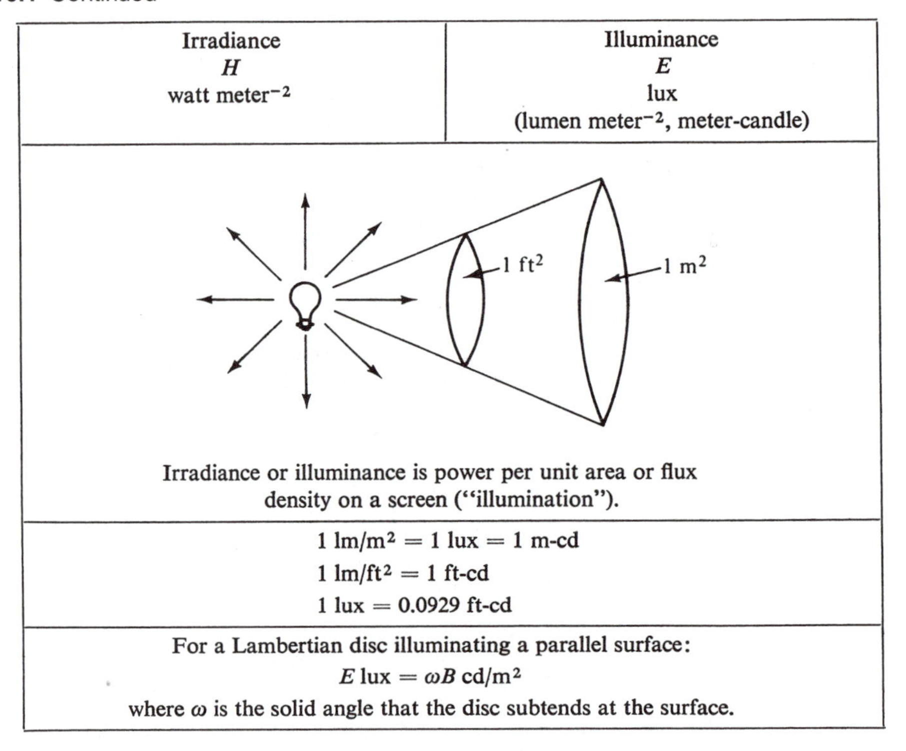

Irradiance *H* watt meter^{-2}	Illuminance *E* lux (lumen meter^{-2}, meter-candle)
Irradiance or illuminance is power per unit area or flux density on a screen ("illumination").	
1 lm/m^2 = 1 lux = 1 m-cd 1 lm/ft^2 = 1 ft-cd 1 lux = 0.0929 ft-cd	
For a Lambertian disc illuminating a parallel surface: E lux = ωB cd/m^2 where ω is the solid angle that the disc subtends at the surface.	

called "brightness"; the unit for B used to be the *lambert*. An important part of the definition of N or B is the specification of the area. N for example should be the same as J per unit area, but we use the normal or projected area dS_n in the relation

$$N = J/dS_n$$

since the energy received at dS' in Figure 10.2 depends not only on the size of dS but also on its orientation. This can also be written as

$$N = J/dS \cos\theta = dP/d\omega\, dS \cos\theta$$

The quantity J, being power per unit solid angle radiated in a direction specified by θ, should also have an angular dependence and be written as $J(\theta)$. If the energy supplied by dS is being received from another source and retransmitted isotropically—that is, dS is a *perfect diffuser*—then $J(\theta)$ is related to its normal component by the equation

$$J(\theta) = J_n \cos\theta$$

This is known as *Lambert's law*; a surface that satisfies this condition is described as *Lambertian*. (Such a surface is equivalent to a collection of point sources radiating isotropically). From this

$$N = J_n/dS$$

which indicates that the apparent brightness of a Lambertian surface is everywhere constant. The sun, for example, has the appearance of a disk of constant brightness. For a Lambertian surface only, there is a relation between radiant emittance W and radiance N which we obtain by considering a plane source radiating dP watts per unit area into a solid angle $d\omega$ (Figure 10.3). Using

$$dP = J(\theta)d\omega$$

then

$$P = \int J(\theta)d\omega = J_n \int \cos\theta d\theta \tag{365}$$

The solid angle $d\omega$ is enclosed by the intersection of the two cones in the figure with the surface of a unit sphere, or

$$d\omega = 2\pi \sin\theta\, d\theta$$

so that the flux is

$$P = 2\pi J_n \int_0^{\pi/2} \sin\theta \cos\theta d\theta = \pi J_n \tag{366}$$

Dividing by the surface area normal to the flux, we obtain

$$W = \pi N_n \tag{367}$$

The photometric equivalent is then

$$L = \pi B_n \tag{368}$$

The factor of π was regarded as a nuisance; to eliminate it, the lambert (mentioned earlier) was introduced. When $L = 1$ lumen/cm^2, then $B = 1$ lambert. In addition, there was the *meter-lambert* (1 lm/m^2) and the *foot-lambert* (1 lm/ft^2), but these (and many more non-SI) units have fortunately disappeared.

The last pair of entries in the table permit a specification of the amount of energy received from a source by the surface dS'. *Irradiance* H (watt/m) is the radiometric

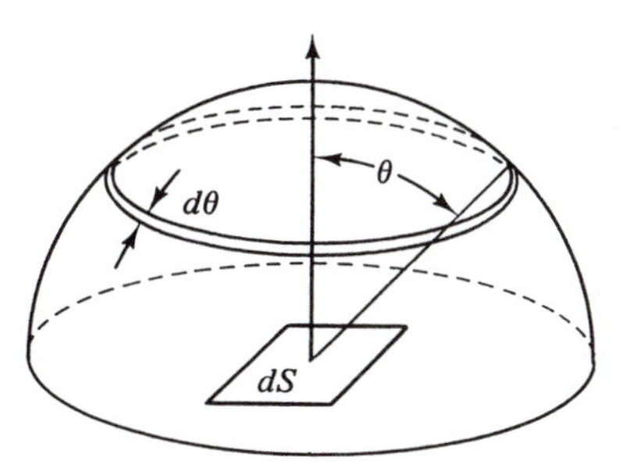

Fig. 10.3

quantity; *illuminance* E, whose SI unit is the *lux*, is the photometric equivalent. Figure 10.4 shows a source, a radiating disk and a receiving surface dS' normal to the z-axis. We wish to find the connection between the luminance B of the radiating disk and the illuminance E at the receiving surface. The flux in the direction shown from an elementary area on the disk is

$$dF = B\,d\omega\cos\theta\,dS \tag{369}$$

where the solid angle subtended by dS' at the radiating point P will be

$$d\omega = \frac{dS'\cos\theta}{r^2} \tag{370}$$

and the radiation comes from the element

$$dS = 2\pi\rho\,d\rho \tag{371}$$

Then

$$F = 2\pi B\int dS'\int_0^R \frac{\cos^2\theta}{r^2}\rho\,d\rho = 2\pi BS'\int_0^R \frac{\cos^2\theta}{\rho^2 + D^2}\rho\,d\rho \tag{372}$$

assuming that B is constant across the disk. Substituting for $\cos\theta$ and integrating gives

$$E = \pi BS'\frac{R^2}{R^2 + D^2} \tag{373}$$

From the definition of illuminance

$$E = \frac{dF}{dS'} = \frac{\pi BR^2}{R^2 + D^2} = \frac{BS}{r^2} \tag{374}$$

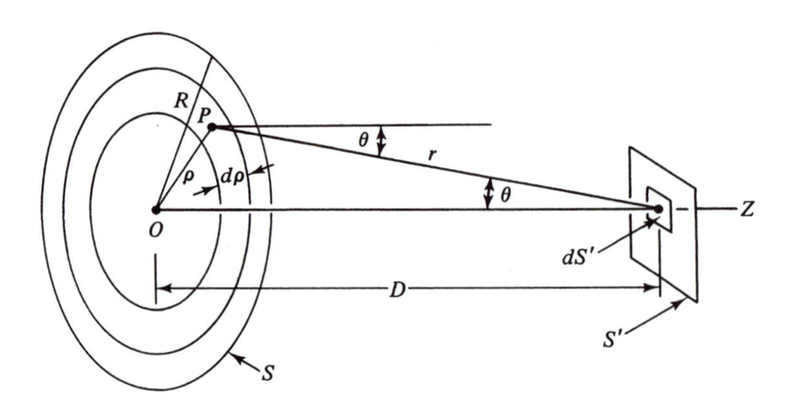

Fig. 10.4

where S is the area of the disk. When dS' is far enough away, Problem 107 indicates that we may finally write

$$E = \omega B \tag{375}$$

which is the desired relation. We should also note that a source radiating 4π lumens of flux in a uniform manner will be providing 1 lm/sr or 1 candela. Hence, 4π or 12.57 lumens of radiant flux from an isotropic source means that $I = 1$ cd. For example, a typical 100-watt light bulb provides about 1,700 lumens of visible energy. If the distribution were uniform, the luminous intensity I would be about 135 cd (it was once customary to call this a 135-candlepower source). This may be compared to a flashlight bulb (1 candlepower) or an arc lamp (10,000 candlepower).

PROBLEM 108

A lens with a diameter of 1.25 in. and a focal length of 2.0 in. projects the image of a lamp capable of producing 3,000 cd/cm^2. Find the illuminance E in lm/ft^2 (footcandles) on a screen 20 ft from the lens.

PROBLEM 109

(a) A lamp of 100 cd illuminates a book 4 m directly below it. Find the illuminance.

(b) A small-aperture spherical mirror is placed above the lamp at its focal distance of 3 m. The reflection coefficient of the mirror is 0.9. Find the increase in illuminance on the book.

10.2 PROJECTION SYSTEMS

The purpose of a projection system is to produce a large, bright, aberration-free image of a transparent object such as a slide, microfiche, or viewgraph. Figure 10.5 shows the essential parts of a typical unit: a microfiche reader that uses 16-mm motion film and projects it onto the rear of a translucent screen. The frame is approximately 0.5×0.5 in., and the screen is 12×12 in., giving a magnification of 240×. The light comes from a lamp with an output of 4,500 lumen, consuming 150 watts at 120 volts. This amount of energy is radiated over a full sphere from a plane filament whose dimensions are $3/8 \times 3/8$ in. For the geometry shown, the solid angle subtended by a frame whose area is 0.25 in.^{2} at a distance of 3.5 in. from the bulb is less than 0.1 steradian, which is only about 1% of the 12.5 steradian for a full sphere. There are two ways of getting more of this light to the object. The first is to place a spherical mirror behind the bulb at a distance equal to its radius. If the bulb were truly a point source, nearly half of the rays leaving it would strike the mirror and be reflected back. For the plane filament, an approximate plane image is obtained. If this image falls on the filament, the concentration of energy in a small space would raise the temperature enough to shorten the filament life. Therefore, the image is arranged to fall next to the filament,

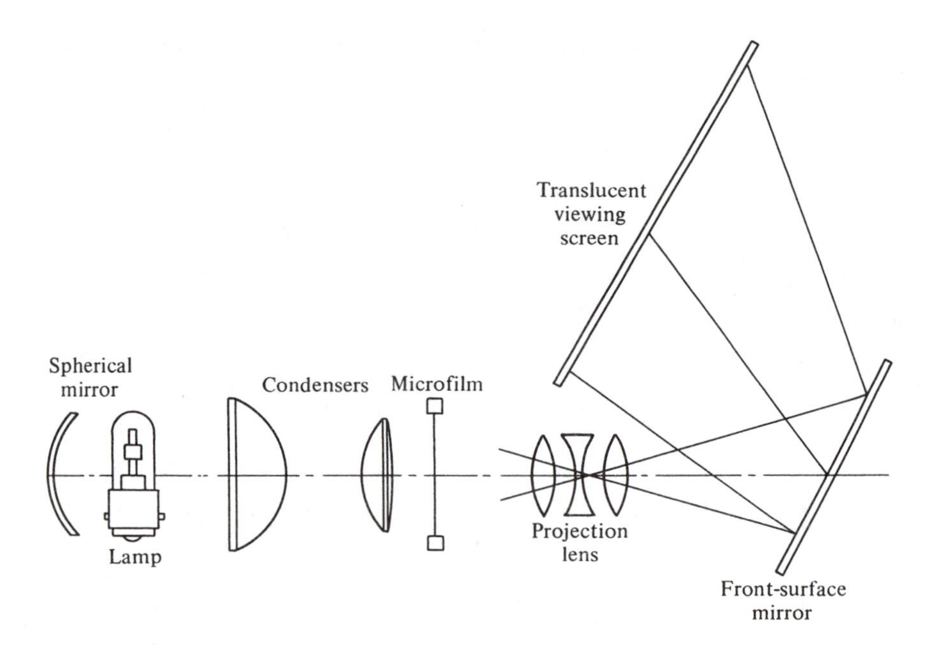

Fig. 10.5

and the result is to double the size of the source, which gives more uniform illumination on the rearview screen. The other improvement shown is the use of a pair of *condenser* lenses. This double condenser takes a beam that is radiated over approximately 2π steradians and decreases the solid angle to match the size of the transparent object. Typically, the condenser has two plano convex lenses with the curved surfaces facing each other, and with a spacing equal to their average focal length. As shown previously, this arrangement reduces chromatic aberration. One or both components may be aspherics; condensers are inexpensive plastic lenses. They can be molded accurately to complex shapes, and they need not be too elaborate in design because reasonable amounts of geometric aberration can be tolerated.

PROBLEM 110

The condenser lenses in Figure 10.6 are spherical, with the following parameters:

r	t'	n'
∞		
	0.624	1.518
−1.087		
	0.985	1.000
1.102		
	0.370	1.523
−6.693		

The filament is 1.0 in. to the left of the condenser. Find the size and position of its image.

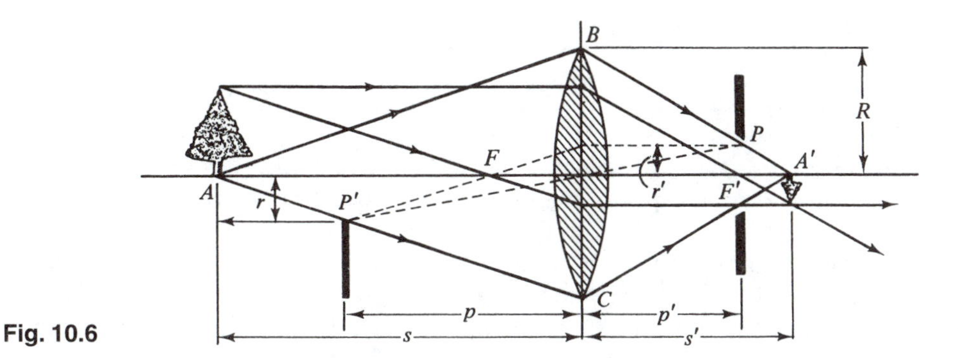

Fig. 10.6

The solution to Problem 110 shows that the image of the filament will be inverted and about 2.6 in. to the right of the condenser, with a magnification of about 1.4. The microfilm object, however, is located at a much closer distance, so that the filament combined with its image as produced by the mirror has a total image that falls somewhere past the left-hand focal point of the projection lens. The cone of light leaving the condenser has a diameter of about 5/8 in. when it passes through the transparent object; therefore it completely illuminates it. The reasoning behind the position arranged for the image of the filament is to prevent the projection lens from producing a second image of the filament at the screen. This would have a disastrous effect on the legibility of the text on the screen. Another consideration is the diameter of the cone as it strikes the projection lens. It is necessary to ensure that it is large enough to use the full aperture of the lens, but not so large as to waste light energy and require a bigger projection lamp than necessary.

Let us digress to examine in more detail energy transmission efficiency. Every optical system has some limiting aperture. Telescopes, for example, have a large, expensive objective and a small eyepiece; the objective is the limiting component because of its cost and weight, and we shall now show that its size is also a limitation—the eyepiece plays no role in this feature of the design. A camera lens has an adjustable opening called an *aperture* or *stop*. Figure 10.6 shows an object—the tree—and a metal plate with an opening whose lower edge, labeled P, serves as a stop on the image side of the lens. Choose an object point A on the axis. The limiting rays are those that go to B and C, and the ray from B to the image point A' just grazes the edge of the stop at P. The left-hand stop in this figure is not real—it is the image of the metal plate lying between the tree and the lens; in particular, P and P' are object and image. We shall now show that the image of the stop in object space has the same limiting effect there as does the stop itself in image space. Similar triangles show that

$$\frac{r'}{s' - p'} = \frac{R}{s'} \tag{376}$$

where r' is the stop radius and R is the radius of the lens (the distance from axis to edge).

PROBLEM 111

Using the Gauss lens equation, show that an equation like (376) is satisfied in object space.

Since the similar triangles in object space are established by the rays that just graze the image of the stop, it equally well establishes the light-gathering ability of the lens. Note that the ray from the top of the tree that passes through F and F' is similarly limited by either the stop or its image. If the stop were moved a little farther to the right, then the lens takes over as the limiting aperture. Whatever establishes the limit—lens or stop—is called the *entrance pupil*. For a more complex system having several lenses and stops, we must find the image of each element in object space and determine which one acts as the entrance pupil. In a similar way, we define an *exit pupil*. For an object to the left of A, the rays forming the image are limited by the stop, so that the stop is the entrance pupil and its image is the exit pupil. For rays starting to the right of A, the lens is both exit and entrance pupil.

PROBLEM 112

A telescope has the following parameters:

	Focal length (cm)	*Diameter* (cm)
Objective	100.0	10.0
Eyepiece	5.0	2.0

If the two lenses are separated by a distance of 105.0 cm, show that the objective serves as the entrance pupil and that its image is the exit pupil. Find the position and size of the exit pupil. This afocal arrangement is very common for telescopes and shows that the objective is the limiting element not only physically (size, weight, cost, and so on), but optically as well.

Light energy passing through a system is conserved if we neglect losses due to absorption or reflection. Consider an area dS normal to the axis, radiating dF lumens into a solid angle $d\omega$. Let the light strike the entrance pupil of a lens at an angle α to the axis (Figure 10.7), intercepting an elementary area when r is large and $d\alpha$ is small, given by

$$dA = (r\, d\alpha)\,(r \sin \alpha\, d\phi)$$

Using

$$dF = B\, dS \cos \alpha\, d\omega$$

then

$$dF = B\, dS \cos \alpha\, dA/r^2 = B\, dS \cos \alpha \sin \alpha\, d\alpha\, d\phi$$

or

$$dF = B\, dS\, d\phi\, d\left(\frac{\sin^2 \alpha}{2}\right) \tag{377}$$

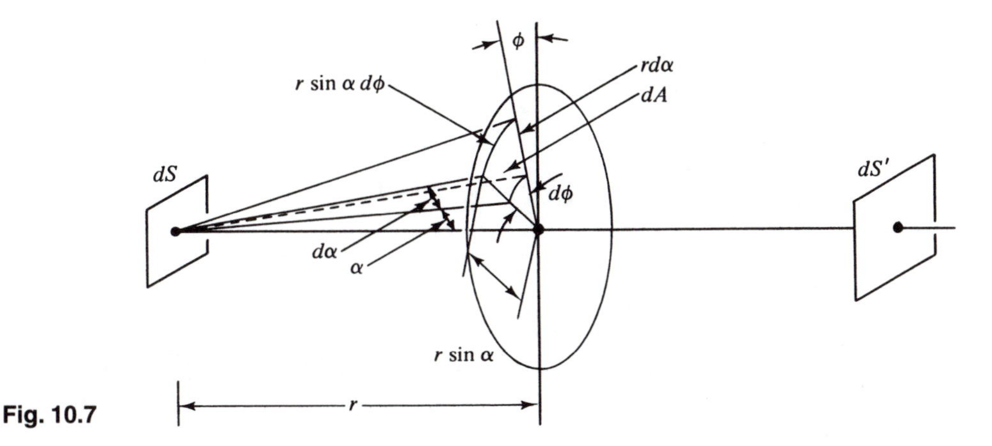

Fig. 10.7

If an image of area dS' is formed to the right of the lens, conservation of energy requires that

$$B\,dS\,d\phi\,d(\sin^2 \alpha) = B'\,dS'\,d\phi'\,d(\sin^2 \alpha') \tag{378}$$

where the primed quantities refer to the image. For perfect imaging of off-axis points, the Abbe sine condition of Chapter 3 must be satisfied. This gives

$$(xn \sin \alpha)^2 = (x'n' \sin \alpha')^2$$

Since the magnification of the area is

$$dS'/dS = (x'/x)^2$$

the sine condition is equivalent to

$$dS\, n^2 \sin^2 \alpha = dS'\, n'^2 \sin^2 \alpha'$$

Putting this into (378), we obtain

$$\frac{B}{n^2} d\phi = \frac{B'}{n'^2} d\phi' \tag{379}$$

For a system with axial symmetry, the angle ϕ remains constant. If object and image are in the same medium, then (379) reduces to

$$B = B' \tag{380}$$

Thus, conservation of energy implies conservation of brightness, subject to the above restrictions. At first sight, this is surprising. If we take the image of a bulb projected onto a screen and reduce it, it appears brighter. What we should really do is observe the image *directly*—for example, with zoom binoculars. Reducing the magnification simultaneously reduces the area dS' and increases $d\omega'$ as subtended at the eye. This is shown in Figure 10.8; if the image size is decreased from dS'_1 to dS'_2, the solid angle is increased from $d\omega'_1$ to $d\omega'_2$, but the product $dS'd\omega'$ stays the same by (380). These

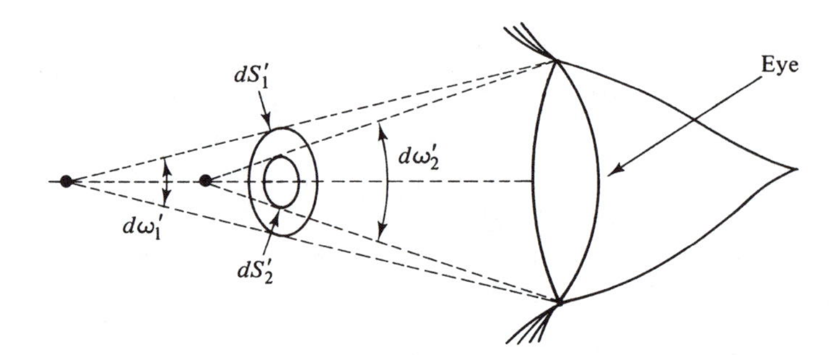

Fig. 10.8

arguments do not apply to images observed on a screen, for then the light is scattered in all directions, and the apparent paradox is resolved.

Returning now to the microfilm reader, the design calls for the image of the filament to be located near the entrance pupil of the projection lens, and—as mentioned previously—it should be of approximately the same size. This image will therefore be formed within the lens body where it does no harm. It has been found empirically that the best results are obtained when the image of the filament lies very close to the last surface of the projection lens. This spreads the light more uniformly across the microfilm plane without making the light cone too large to pass through the entrance pupil. This latter requirement will be recognized as the optical equivalent of impedance matching in a circuit to obtain maximum power transfer. Since the filament plus its image from the mirror is a rectangle with a diameter of about 7/8 in., the image has a diagonal of 1.2 in. and a large and expensive lens is called for. The actual instrument on which this discussion is based used a lens for which the entrance pupil diameter was only 0.58 in., and the mismatch was large, being a factor of 4 in area. To find the illuminance E at the viewing screen and determine the effect of this mismatch on the performance of the system, we first determine the brightness B at the entrance pupil. The front surface of the condenser system will receive a portion of the radiation from the 4,500-lumen bulb as determined by the solid angle it subtends at the source. Given that the filament-to-condenser distance is about 1.2 in. and the condenser radius is about 1 in., the solid angle is then $(1/1.4)^2\pi$, or about 0.7 π, and

$$\mathrm{B} = 2 \times 4500\ (0.7\ \pi/4\ \pi) = 1600\ \mathrm{lm/sr/cm^2}$$

Assuming about a 15% loss of illumination in the condenser and 5% in the projection lens reduces this to 1,300 cd/cm^2. The projection screen is 40 in. from the lens so that the illuminance it receives is

$$E = B\omega = 1300\ (\pi\ 0.29^2/40^2) = 0.15\ \mathrm{lm/cm^2}$$

For convenience, we convert this as follows:

$$(0.15\ \mathrm{lm/cm^2})(929\ \mathrm{cm^2/ft^2}) = 200\ \mathrm{lm/ft^2} = 200\ \text{ft-candles}$$

Measurements gave 23 ft-candles at the center and 19 ft-candles at the corners of the screen. The loss of a further factor of 4 due to the mismatch at the entrance pupil brings the calculation and the measurement into moderate agreement. The falloff of illuminance from center to corner is also significant. It is known that the human eye can tolerate a variation of about 50% without serious objection. Our calculations are valid only for the center of the screen. We now extend them to a region off the axis. Figure 10.9 shows point A where the axis crosses the screen, and the solid angle subtended there by the entrance pupil is the area divided by the square of the distance. For point B, the solid angle is due to the disk with edges at F and G, and the area is decreased by the factor cos ψ. The distance to the screen is increased by the same factor, and this appears in the solid angle as $\cos^2 \psi$. Since illumination is measured on a surface normal to the direction of propagation, there is a further cos ψ factor at the screen. The net result of all this is a falloff from center to edge of the screen by an amount that varies as $\cos^4 \psi$. This variation is in agreement with the measurements just cited. The screen is about 12 in. in width, and the associated cosine has a value of about 0.99, with a fourth-power of about 0.95.

10.3 PERISCOPES

The ideas introduced in the previous section can be applied to long-distance viewing. Start with the thin lens of Figure 10.10, for which $f' = 1$. It produces an inverted image for which $m = -1$ when the object is 2 units to the left. Another identical lens placed at 2 units to the right of this inverted image acting as an object will have an erect image with $m = +1$ and a location 2 units to the right of the second lens.

PROBLEM 113

Locate the cardinal planes for this pair of lenses and show that the unit planes are at the object and image positions, as they should be.

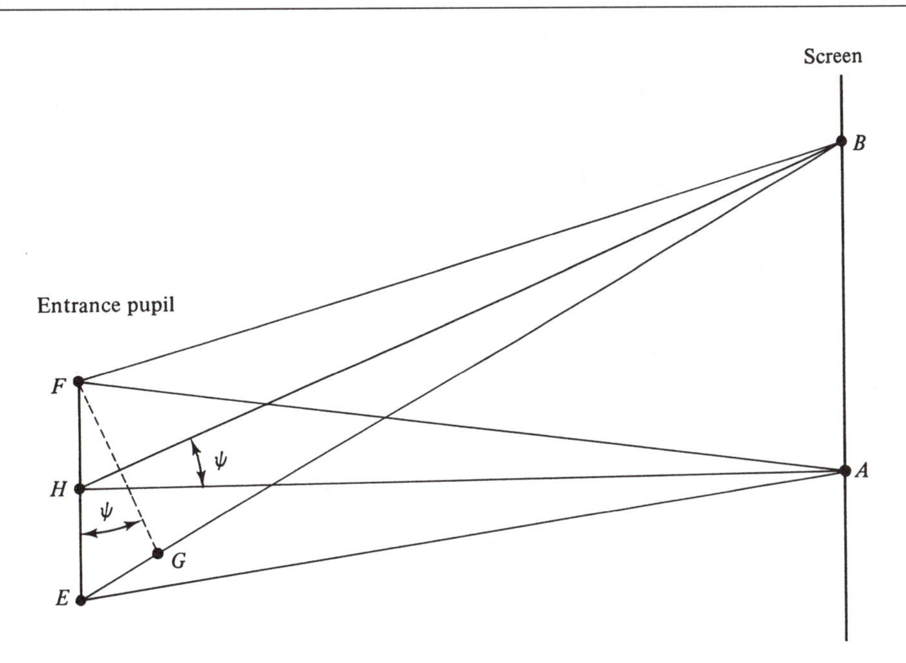

Fig. 10.9

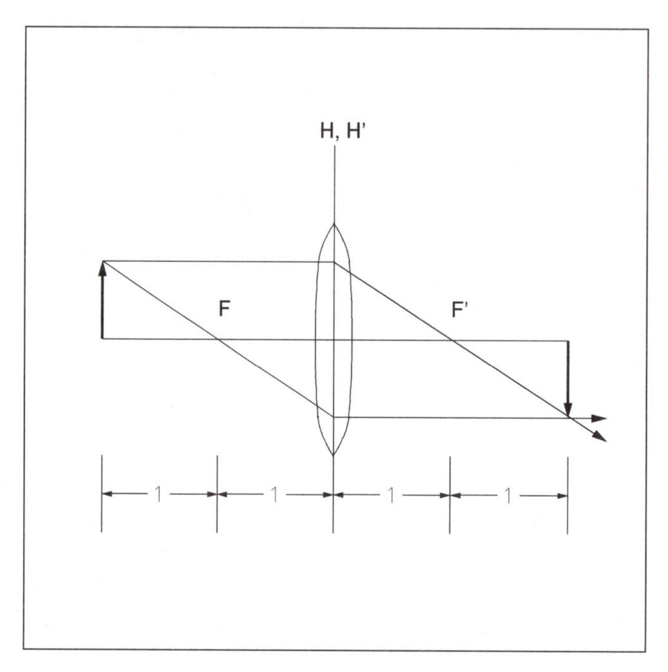

Fig. 10.10

Now we take into account the size of these two lenses; let them be 2 units in diameter (Figure 10.11). The three rays we normally trace are shown leaving P. These represent the complete fan from P that can reach the first lens. Two of these rays (rays 1 and 2), however, will not get to the second lens, and ray 3 just grazes the edge. Thus, no image of P is created at H′. For an object point at A, halfway between P and the axis, we see that ray 4 misses the second lens, ray 5 just grazes the edge, and ray 6 is

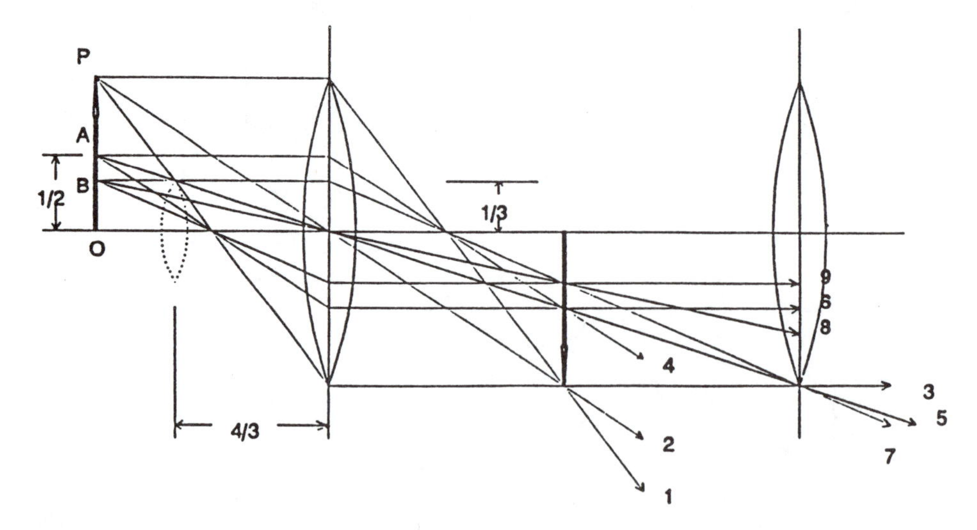

Fig. 10.11

refracted. There is partial transmission or *vignetting*. For point B, at x = 1/3, ray 7 just grazes the lens, but all the others get through. Figure 10.11 also shows the image of the second lens as created by the first lens. By Gauss' equation, this image is 4/3 unit to the left of the first lens and it has a half-height of 1/3. It forms the entrance pupil of the system and is large enough to accept all rays needed to form an image of an object of the size of OB; larger objects will be partially blocked. And when the rays get through this image they also get through the second lens, so that it acts as the exit pupil. The same physical entity serves as both entrance- and exit-limiting features.

To enlarge the entrance pupil, we add a third, identical lens at the center; this is the *field lens*. Its image in the first lens coincides with OP and the entrance pupil now accepts the full object. This arrangement is a *periscope*, which is an instrument for viewing objects at a remote location. In this example, the field lens has no effect on the image size.

PROBLEM 114

Use the paraxial matrices to confirm this statement.

If more lenses are added to the system, all equally spaced, we have an *optical relay*. Mount this system in a tube, with mirrors set in at an angle of 45° near each end. Then we have specified the type of periscope used in a submarine. Suppose you look down a long tube the diameter of which just accepts these lenses, but you're looking before the lenses are inserted. At the far end, you can see a very small section of the landscape. After constructing the relay, however, the change in the entrance pupil gives the impression of looking through an aperture in a metal plate with a diameter of 2 units so that, if the eye is 2 units away, then the field of view has an angular size of about 90°. A new building at the University of Minnesota has a lobby at ground level, and the rest of the building is underground. People with offices well below the street have large openings in their walls that are connected with periscopes whose upper ends are high above the structure. When they look through these "windows," they see the outdoors at full size.

PROBLEM 115

An afocal telescope has an objective with a focal length of 60 cm and a diameter of 5 cm. The eyepiece consists of two identical lenses for which the focal length is 4 cm, the diameter is 2.5 cm, and the spacing is 3 cm.

(a) Demonstrate that the objective is the entrance pupil and its image is the exit pupil.

(b) Show that the location of this image gives an *eye relief* of about 2.5 mm.

PROBLEM 116

It was explained in connection with Figure 7.12 that lens speed, or f#, is defined for objects at infinity. If the object is closer, then the effective speed changes because the solid angle subtended at the image point decreases. Show that the new image point moves away from the lens by an amount determined by the magnification, and that the effective speed is decreased by a factor $(1 + m)$, where m is the magnitude of the magnification.

10.4 DEPTH OF FIELD

An important practical aspect in photography is the concept of *depth of field*. This measures the ability of a lens to produce images that are in focus not only for the primary object but over a distance in front and in back of it as well. To give this term a quantitative definition, consider Figure 10.12, which shows an object at P and its image at P′. There is a blur circle of radius ε in the image plane created by the rays coming from points A and B. Let the location of the images A′ and B′ determine the largest acceptable size of the blur circle. If P is at a distance s from the geometric center of the lens and P′ is at a distance s', then the Gauss lens equation connecting these quantities with the image space focal length f' is

$$1/f' = -1/s + 1/s'$$

or

$$s' = f's/(f' + s)$$

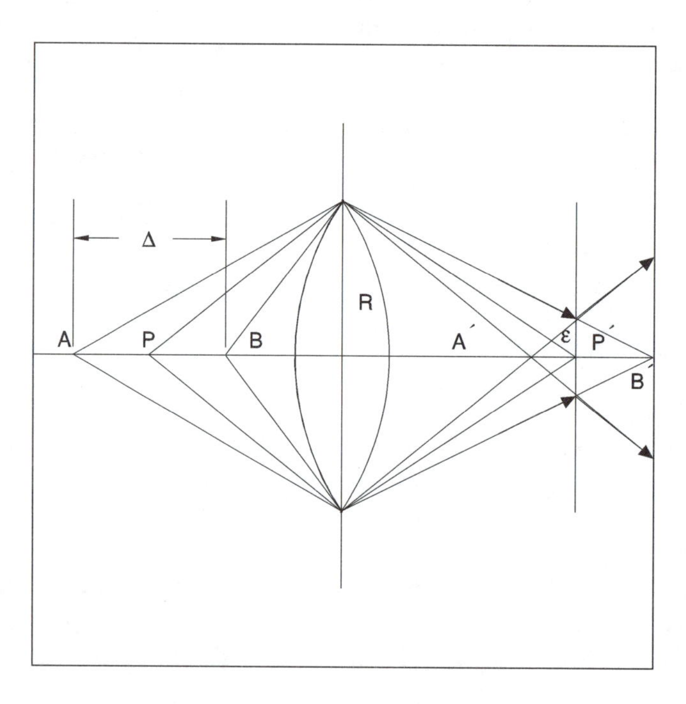

Fig. 10.12

Let ds denote the magnitude of the small distance from A to P or B to P (to a reasonable approximation, these distances will be the same), and let ds' be the corresponding quantities on the image side. Then differentiating

$$ds' = f'^2 ds/(f' + s)^2$$

Approximately, by similar triangles

$$\varepsilon/ds' = R/s'$$

where R is the radius of the lens. Then

$$ds = (f' + s)^2 ds'/f'^2 = (s + f')^2 \, \varepsilon s'/f'^2 R$$

But usually s is much larger than f', and s' is about the same as f' so that

$$ds = (\varepsilon s'/R)(s/f')^2 = \varepsilon s^2/Rf'$$

The depth of field Δ is $2ds$, as Figure 10.12 shows, and the speed of the lens is expressed as an f-number, where $f\# = f'/2R$, giving

$$\Delta = 4 \text{ f\#} \, \varepsilon \, (\text{s/f}')^2$$

(Do not confuse this quantity with the *depth of focus* A′B′.) As an example, consider a 35 mm camera with a 2.8-cm wide-angle lens whose speed is f/4. The resolution of 35 mm color film is usually about 10 μm. For an object at a distance of 10 cm, the depth of field will then be

$$\Delta = 4 \times 4 \times (10/2.8)^2 \times 10^{-5} = 2 \times 10^{-3} = 2 \text{ mm}$$

As another example, A. Cox, in the book *Photographic Optics* (Focal Press, 1966), gives a series of depth-of-field tables. For a 5-cm, f/4 lens with an object at 10 ft, the region for which the image will be in focus is 8 to 13 ft. Cox stipulates that the blur circle is about 25 μm or 0.001 in. Then

$$\Delta = 4 \times 4 \times (120/2)^2 \times 0.001 = 58 \text{ in} = 5 \text{ ft}$$

which checks nicely with Cox's tables.

A

B

C

G

H

I

J

K

L

M

N

O

P

Q

R

S

T

U

V

W

Z